Are You Driving Your Children To Drink?

Coping With Teenage Alcohol and Drug Abuse

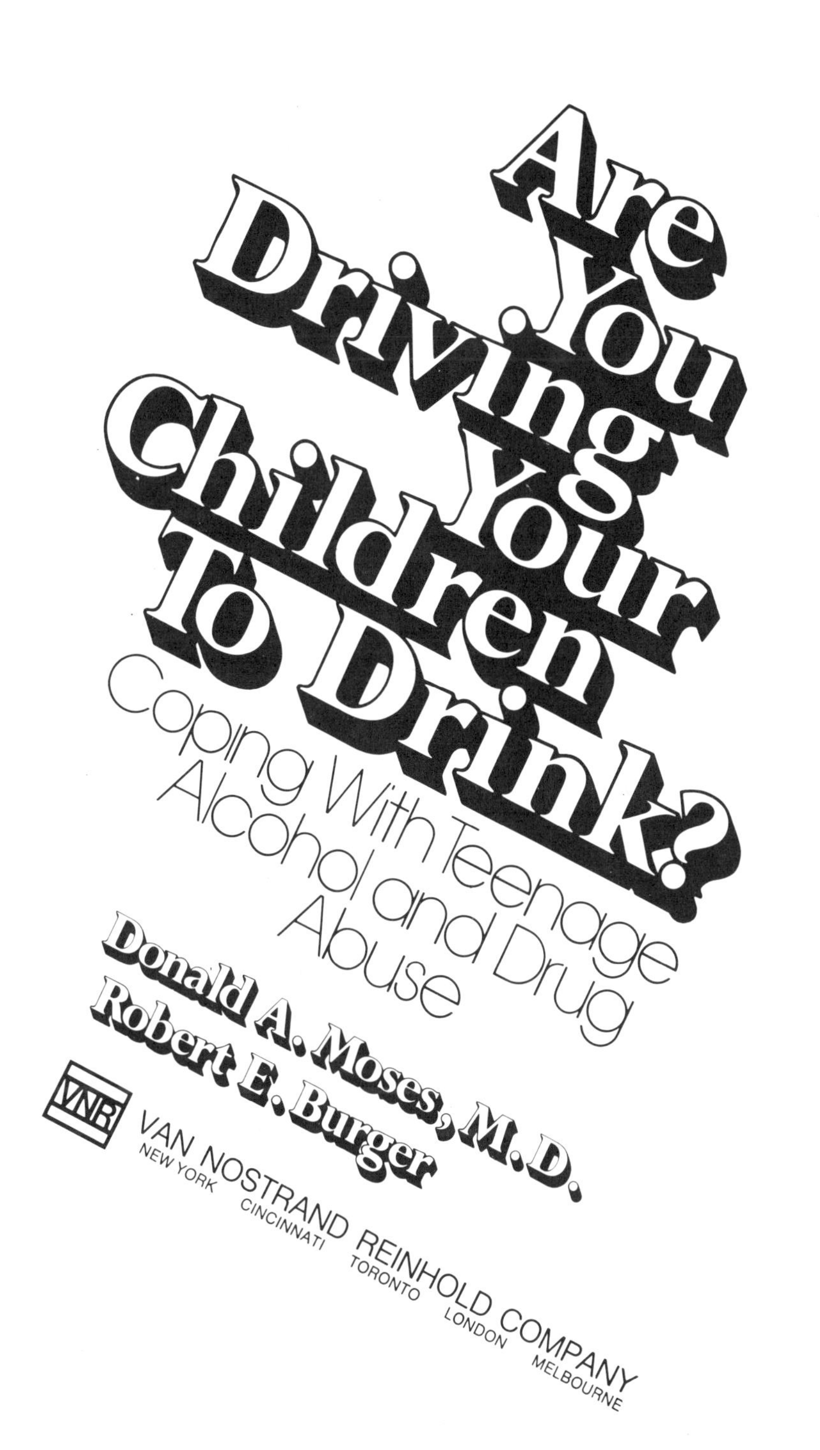
Are You Driving Your Children To Drink?
Coping With Teenage Alcohol and Drug Abuse
Donald A. Moses, M.D.
Robert E. Burger
VNR
VAN NOSTRAND REINHOLD COMPANY
NEW YORK CINCINNATI TORONTO LONDON MELBOURNE

Van Nostrand Reinhold Company Regional Offices:
New York Cincinnati Chicago Milbrae Dallas

Van Nostrand Reinhold Company International Offices:
London Toronto Melbourne

Library of Congress Catalog Card Number: 75-23097
ISBN: 0-442-25583-7

Manufactured in the United States of America

Published by Van Nostrand Reinhold Company
450 West 33rd Street, New York, N.Y. 10001

Published simultaneously in Canada by Van Nostrand Reinhold Ltd.

15 14 13 12 11 10 9 8 7 6 5 4 3 2 1

Library of Congress Cataloging in Publication Data

Moses, Donald A
Are you driving your children to drink?

Bibliography: p.
Includes index.
1. Drugs and youth. 2. Alcohol and youth. 3. Drug abuse—Treatment. 4. Alcoholism—Treatment. I. Burger, Robert E., joint author. II. Title.
HV5801.M67 362.7'8'29 75-23097
ISBN 0-442-25583-7

to
SARAH
my wife
RICHARD and ERIK
my sons

And to those youngsters who had the insight to begin, the courage to continue and the perserverance to change; giving me the knowledge that made this book possible.

Preface

This book presents a new approach to the understanding and treatment of the drug and alcohol problems of adolescents. The concept we propose—which is essentially that the psychological interaction of parents and children is at the heart of the problem—is based on several years of work with adolescents in a New York State day-care center. The range of cases studied is, however, broadly representative of the United States population at large. Likewise, the psychological models which we use to analyze the drug-abuse syndrome are not fads of the moment, but are basic to well-established counseling techniques.

We use the word "drugs" in its widest sense, involving psychological *or* physical habituation. Alcohol is, of course, the most commonly used and abused drug in our culture; and largely for this reason it is our most dangerous drug. But in common parlance "drugs" tends to mean marijuana, morphine, cocaine, and so forth—those drugs which are not taken as food or drink but ingested in other ways. Since we cannot change the language by edict to reflect the effects which are common to all drugs, we are forced to say "drugs and alcohol" in specifying the basic consciousness-altering drugs. The reader will, we hope, bear in mind that when we refer to "drugs" alone we usually mean to include alcohol (and even nicotine and caffeine) under that rubric. The treatment of alcohol abuse, however, is often an easier matter than the treatment of "hard drug" abuse—in a general way, because of the "oral" nature of alcohol abuse. Other drugs also have specific psychological importance. In any case, our point is that alcohol and drug abuse should be thought of not as the cause, but as the result, of maladjustive behavior.

There are cycles in the popularity of certain drugs as there are in everything else. In recent times, rebellious youth has turned to alcohol as a release from the pressures of the family and society. But the "drug of choice" is not so important as *the reason for using it*. If the cycle swings back again to cocaine or to marijuana, the causes will remain the same. It is the purpose of this book to get at the causes: to show parents where they may have unwittingly planted the seeds of frustration and rebellion that have literally driven their children to drugs and drink. Once the basic mechanisms of parent-child relationships are understood, we can intelligently proceed to treat the roots of the drug problem instead of the branches. Best of all, we can begin to grasp some of the depth and sensitivity of what it means to be a parent, and to develop a fuller awareness of our responsibilities to our children from the moment they are born.

DONALD A. MOSES, M.D.
ROBERT E. BURGER

Contents

PART THREE: HEALING

Therapy takes many forms, but it must break a vicious circle to succeed.

PART FOUR: THE PUBLIC PROBLEM

Let's have an end to public moralizing and find a way to the understanding of individual needs.

Are You Driving Your Children To Drink?
Coping With Teenage Alcohol and Drug Abuse

The Language of This Book

Before we attempt a new approach to a subject, we must all be sure we are talking the same language. Some of the language in the following pages will be new to many readers; what is worse, many others may have popularized misconceptions of the meanings of certain words. Whatever you may have read before, we must agree on the following understanding of certain key psychological terms. We suggest that you scan this theoretical presentation briefly—not study it. Then refer to it whenever the context of the discussion does not make the meaning of a psychological term clear.

When Freud originally created a theoretical structure of the psyche (those aspects of the mind involved with thought and feeling), he divided it into three areas:

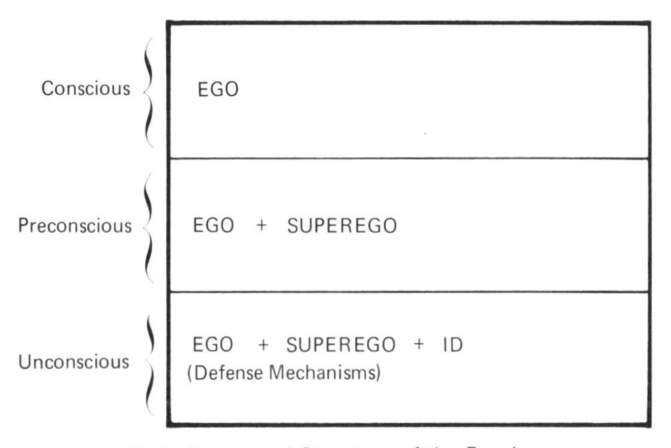

Early Concept of Structure of the Psyche

(1) *Conscious (Cs)*—That part of the mind of which we are immediately aware. "It is daytime." "That is a tree." "I am happy." The conscious is a relatively small part of the psyche, as we can be aware of relatively little at any moment.

(2) *Preconscious (Pcs)*—That part of the mind whose memories can easily be brought to awareness with a moment of thought. "Last week we went to the zoo." "His name is, uh, George."

(3) *Unconscious (UCS)*—Sometimes referred to as the subconscious, this is by far the largest part of the psyche. It is composed of forgotten memories, "repressed" desires, and forbidden wishes. All of the impingements on the psyche of the early years of life are stored here: the thoughts, feelings and memories that cannot be brought to awareness no matter how hard one tries. The forbidden instincts that man must control in society are largely confined to the unconscious.

It was a number of years before Freud decided that this classification was inadequate and formulated a new theoretical structure, which merges with his original.

	Conscious	Unconscious
EGO	That Which Deals with the World Intelligence, Assimilation, Synthesis, Memory, Imagination Perception, Sense of Self, Sense of Identity, Pride, Shame	That Which Deals with the Self Repressed Memories and Feelings Defense Mechanisms
SUPEREGO	Conscience (Concept of Right and Wrong) Guilt	Fear of Punishment, Disapproval, Rejection from Others Guilt
ID		Drives: Libido, Aggression

Later Concept of Structure of the Psyche

(1) *Id*—The basic instincts of survival, leading to the need to gratify specific urges. The id is divided into *libidinous drives* (erogenous urges such as eating, elimination and sexual activity) and *aggressive drives* (to obtain food and defend family and territory, as distinguished by Konrad Lorenz). Many orthodox Freudians do not separate these two general categories of drives, believing that aggression stems from frustration of libidinous

wishes. The id is composed of energy that is usually surging for expression and constantly being frustrated. The entire id "resides" in the unconscious, and the only awareness that we have of it is by its derivatives (the results of its urges). We are unable to see the wind but are made aware of its presence by the movement of the leaves. It is because of the *frustration* of the drives in the id that conflicts arise.

(2) *Ego*—Divided between the conscious and unconscious is that "executive of the personality" which allows us to deal with reality. The ego is the intermediary between the self and the world on the one hand, and the self and the unconscious on the other. The conscious functions of the ego include: intelligence (thought), perception (awareness and understanding of what enters through our senses), assimilation and synthesis (the process where old ideas become new thoughts), memory, imagination, sense of self, sense of identity, pride, shame, etc. The unconscious functions of the ego are the defense mechanisms that hold the id impulses in check and keep unwanted memories from the conscious.

In order for the ego to keep the unconscious form from being conscious, it employs certain defense mechanisms:

(a) *Repression*—forgetting unpleasant memories without trying.
(b) *Suppression*—forgetting unpleasant memories on purpose.
(c) *Sublimation*—bending an unwanted urge to positive action. A potential ax murderer becomes a butcher; a voyeur becomes a photographer.
(d) *Reaction-formation*—Reversing one's true feelings. "If I show my hatred to baby brother, I shall be punished; I had better love him."
(e) *Projection*—Placing unwanted, unconscious feelings from the self onto another. The source of paranoia and hallucinations. "I hate him. I mustn't hate him. He hates *me*." Note that projection does not mean anticipating how others would feel based on one's own feelings.
(f) *Displacement*—The redirecting of unwanted feelings from one person or thing to another. The source of phobias. "I am afraid of my father. I cannot be afraid of him because I love him. I am afraid of male teachers."
(g) *Denial*—One of the most infantile defenses: not admitting unpleasantness to oneself. "What, my child take drugs? Impossible." "What, me overdose or get arrested? Not me."
(h) *Rationalization*—A more sophisticated form of denial: Giving oneself an intelligent, rational reason for irrational thoughts or behavior. "I only used drugs because my friends did." "My child only used drugs because he was with a bad crowd."
(i) *Identification with the aggressor*—"If you can't lick 'em, join 'em." A dramatic example was concentration camp victims who began to emulate their tormentors. Commonly seen in children with hostile, violent parents whom they despise, who become hostile and violent themselves.
(j) *Turning anger against the self*—Fear of externalized anger leads the individual to turn it on himself. A specific form of displacement, this mechanism is employed in suicide and other self-destructive behavior. It is common among drug abusers.

(k) *Doing and undoing*—An assistance in the suppression of anger. The mechanism behind obsessive thoughts and compulsive behavior. "If I act I will destroy with my anger. Therefore, I must undo every action." A person turns a light on and off five times to make sure it is out.

(l) *Withdrawal*—An action defense, physically avoiding any anxiety-provoking situation. Fear of failure in school, for example, forces a child into truancy.

(m) *Behavioral defenses*—Often mislabeled "acting out," which applies only to therapeutic situations. A behavioral defense is a repetitive action which masks any anxiety.

Defense mechanisms are unconscious processes which are not under the control of the individual. It is the object of psychotherapy to make them conscious, and to allow the unconscious urges against which they defend to come to the surface.

(3) *Superego*—Split off from the ego the superego is the "watchdog" of the self: a Jiminy Cricket to a person's thoughts and actions that acts as a censor. The superego includes the person's conscience, which is the conscious, intelligent guide to right and wrong. It also includes guilt feelings and fear of punishment. Guilt is the feeling that "I did something bad and I should be punished." Often illogical, it can overwhelm the intelligent knowledge that there is really nothing wrong with these actions—and in this sense it is distinct from "conscience." Guilt is frequently seen in areas of sex, where, for example, a person married for fifteen years can still feel guilty about enjoying sexual activity. Just as illogical as guilt is the fear of punishment, of disapproval, or of ridicule. Many adults still question "What will my parents feel?" and worry about an action which intellectually they know is proper. The superego is the source of shame and embarrassment felt in the ego. Like the ego, it too straddles the conscious and the unconscious.

Oral stage—As the infant grows into adult life, he passes through a number of stages of development prior to the eighth year. The earliest is known as the oral stage, and occurs from birth to prior to the second year. In the oral stage the primary erogenous area (that area in which the libidinous wishes predominate) is the mouth. The sucking instinct predominates at first and then is replaced by chewing. During this stage the primary psychic drive is to be nurtured, mothered and protected. The mother figure is developed, separating her from other people in the infant's world. As everything is experienced in terms of the mouth, the important people in the child's life are *incorporated*, or "taken whole," into the child's psyche. Incorporation is a very important concept to understand as it is a prominent feature in the development of a drug abuser.

Incorporation—Accepting into the psyche the entire person of someone who is important to the child, without differentiation between desirable and undesirable characteristics of that person. A child becomes *like* the parent or other significant person whom he incorporates, including those characteristics which the child may dislike. For example, a child likes a parent's intelligence but dislikes his alcoholism. With the process of incorporation the child strives for intelligence, but also becomes alcoholic. Though normal in infancy, incorporation becomes an abnormality in later years. It is to be differentiated from identification.

Identification—The process wherein the child begins to differentiate those desirable from undesirable characteristics in the parent—rejecting the undesirable and identifying only with the desirable. Identification is a maturation process necessary in the formation of a mature ego and superego.

As the individual is entirely dependent upon the succor of others, any expression of anger on his part is fearful. *Passive-aggressive* behavior develops in fear of showing anger by saying "no."

Passive-aggressive behavior—The expression of hostility through passive means. The mothers says "eat" in a threatening way, and the child refuses to open his or her mouth. In adolescence the mother says "study," and the child watches television. The belief in the ability to command respect with the word "no" has not yet been recognized.

Anal stage—This stage is the second stage of development after the oral stage; the primary erogenous zone is the anus. It is the stage of mastery of the anal sphincter allowing toilet training to commence. It is the first time when a child rather than the parent is asked to withhold voluntarily from giving in to an impulse and to delay gratification. It is the age where the child begins to learn mastery of the self. It is accordingly the age of the first active conflict between the parent and child; hence the name "terrible twos." Socialization and the forerunner of a superego both begin in this stage. Also the word "no" begins to take on meaning and the child can pass, if not threatened, from a passive-aggressive mode to an aggressive mode of behavior.

Phallic stage—The child first discovers the genital area as a source of gratification and comfort. Little children are often seen with their hands on their genitals, and often admonished not to keep them there. There is a recognition of sexual differentiation and the beginning of identification with parents of the same sex. "Penis envy" and "castration anxiety" emerge, and competition with the parent ("phallic striving") begins. (These terms cannot be taken at face value, but they are not pertinent to the theme of this book.)

Oedipal stage—The final tumultuous stage of infantile sexuality, following closely on the phallic phase. The competitive feeling toward the parent leads to a competition for the parent of the opposite sex. Often one hears the five-year-old remark, "When I grow up, (Mommy or Daddy), I'm going to marry you." Accompanying this feeling is the unconscious desire to get rid of the rival, the other parent. As the rival is a person the child loves, admires, and identifies with, severe conflicts arise that are not resolved until the child finds his own mate. Frequently, oedipal conflicts are never completely resolved.

Latency stage—Somewhere between eight and ten years, strong libidinous drives become quiescent. The child becomes free of conflict and anxiety diminishes. Teaching becomes easier, as do home relationships. It may be called the calm before the storm of adolescence.

Other psychological terms will find their best definition in the context of the discussion. With this basic framework as a background, we can now proceed directly to the problem of the emergence of drug and alcohol abuse as a form of adolescent rebellion.

—Part One—

The Roots

Drugs and alcohol have come to middle America, but the home is where they got their start.

—1— Drugs* Come Home

In the past decade the scourge of drug abuse has spread from the ghetto to suburbia, from the cities to the small towns of our nation. It is a subject that is no longer discussed clandestinely in sociological reviews of the life of the poor and minorities. It is debated openly in schools, city councils and churches. Hard drugs, like alcohol, are now part of white, middle class life.

When drug abuse was contained in the ghetto, it was looked upon simply as a matter for the police, and the only treatment that was suggested was the stern hand of the courts. Sociologists claimed that drug abuse was spawned by conditions of a low economic standard of living. When drugs were limited to the ghettos, it was difficult to refute this argument and this "treatment." As the drug problem began to spread, the sociological argument fell into disrepute. Yet this turn of theory was equally illogical because sociological and economic factors play an obvious role in drug use. Different drugs predominate in different socio-economic groups. While marijuana, alcohol, and barbiturates usually are popular in middle class suburbia, heroin, cocaine, and *alcohol* are the most frequently abused in lower socioeconomic communities.

Because for so many years the treatment of drug abuse has been left up to law enforcement agencies, a public attitude has developed about drugs on the model of crime detection and prevention. The police explanation has often been that drug abuse is caused by the pusher, by the criminal element in society forcing drugs upon unsuspecting, innocent youth. The image of the drug pusher in the thirties and forties—a man standing on the street corner as pictured in Tom Lehrer's famous song "The Old Dope Peddler"—persisted into the fifties.

---

*The use of "drug" here, as explained in the preface, also refers to alcohol as used by today's teenager.

More recently, however, evidence has shown that this depiction is grossly misleading. Most of the drugs that are sold in schools are sold by youngsters who are drug abusers. Rarely is a youngster *forced* to take drugs; he usually seeks them out on his own accord.

Further, it was believed that by making the drug laws more stringent and by imposing stiffer penalties for possession and sale, there would be a marked diminution in the use of drugs and an increase in the number of drug abusers seeking treatment. "Penalize the pusher" was the cry. But after two years, the enactment of such laws in New York State appears to be a failure. There has been no significant increase in the number of addicts or abusers who are seeking treatment voluntarily. Furthermore, while superficially it appears that the use of drugs is on the wane, the reality is that drugs are not used as brazenly. Drugs have gone underground. It is no surprise to those of us involved in drug treatment that alcohol has suddenly risen to the top of the list of the most abused drugs at the adolescent level.

Making drugs illegal does not end the craving for the use of drugs. Legal prohibitions merely turned many youngsters away from marijuana and barbiturates and toward the legal "liquid drug," alcohol. The subsequent increase of adolescent alcoholism has created a situation potentially more serious than anything caused by the abuse of other drugs.

A further extension of the law enforcement approach was to stem the flow of heroin from the poppy fields of Turkey. For the past two years heroin has all but dried up on the streets of New York and presumably throughout the rest of the nation. The addict who used to turn to heroin, however, has now turned to methadone, which is readily available legally and otherwise. In the methadone maintenance and detoxification treatment programs, distribution is often not carefully supervised, and a good portion winds up on the street. Stemming the source of supply has not stemmed the craving for opiates or their derivatives.

If the sociological approach and the law enforcement approach have foundered, what line of defense do we have left? In the past, the psychiatric profession, psychologists, and psychiatric social workers have been unsuccessful in treating the drug abuser. Theories which attempted to explain what lies behind the use of drugs have been occasionally helpful, but have proved to be impractical for widespread treatment of addicts. It is within the framework of psychoanalytic theory, nevertheless, that one begins to recognize an underlying current that flows throughout all socioeconomic levels and various types of abuse and addiction.

For the past several years I have been involved as a psychiatric consultant for the drug programs of the Long Island Jewish Hospital in Nassau County, New York. During this time I have interviewed hundreds of youngsters, have had two-score in analytic psychotherapy and in group therapy. These youngsters come from all walks of life, from the very wealthy to the very poor, blacks, whites, Orientals, Puerto Ricans. I have worked in close harmony with other professional and paraprofessional specialists. The conclusions that are set forth in this book are not mine alone, but are shared by social workers from various disciplines, psychologists, social workers, pediatricians, and internists. Most importantly, these conclusions are shared by ex-addicts who have had little or no formal education, no psychological orientation, no Freudian theory, but have learned from the streets, therapeutic communities, and from their experience as drug abusers. *The one factor that we have found common with all drug abusers is a disruptive family setting.* The lack of a stable family, the lack of parental understanding of the needs of the child, the parent's involvement with the child only as part of the self, or, in the extreme, the parent's total lack of involvement with the child—any or all can play leading roles in the formation of the drug abuser.

The "law enforcement" mentality answers that the drug abuser would not be able to obtain drugs if they were not available. The sociologist may say that the abuser would not turn to drugs if society itself were not so drug and comfort oriented. Nonetheless, the question still remains: Why do some children turn to the use of drugs as a consistent source of escape from the world of tensions, stress, and responsibility, while other youngsters, in the same socioeconomic setting, with the same intelligence, in the same school and with the same drives, refrain from drug abuse and become responsible members of the adolescent community?

A "drug abuser," as we use the term, refers to that individual who relies on drugs as a necessary part of his existence. He most likely suffers a psychological habituation, that is, an inability to function on a day-to-day basis, to face the stress and responsibilities of daily life, such as school, social relationships, and family obligations. He may also occasionally develop a physical addiction, in which a daily dose is required for his functioning. But the addict is *not* the youngster who will experiment with different drugs and give them up after a short period of experiencing their effects. Nor is the addict the occasional user whose life remains stable, productive, filled with healthy interpersonal relationships and a goal oriented future. These youngsters, much as those adults who imbibe alcohol at a weekend party but remain functioning, healthy, and related to their daily life, are not the ones who cause the problem to themselves and to society.

Confusion about the causes and cures of drug abuse has caused widespread fear about the subject. It has become a highly charged issue, and emotions in themselves have helped to distort the facts. Into this maelstrom, the shift of adolescent drug abuse from "drugs" to alcohol has come like a rescuing angel for many parents. Is alcohol the way to wean a youngster away from his addiction? Is it less dangerous both legally and pharmacologically?

The most commonly used forms of ethanol among adolescents have been wine and beer, especially the "soda pop wines" which are both sweet and inexpensive. These seem innocent enough. Yet the popular misconception is that alcohol has been *replacing* illegal drugs, when in actuality it is being used in *addition* to other drugs, most commonly in combination with marijuana and the sedatives. Adolescence is also a time of fads, and at the present time drinking is the "in" thing to do. Whatever the causes, alcoholism in youth is rapidly on the increase. All the signs and symptoms of the older alcoholic will manifest themselves in the young in time: intoxication, loss of a sense of purpose, and the physiological symptoms of brain damage, hepatic damage, and neurological damage that once were reserved for the older population. Drugs have come home to the mainstream of America, *yet they were really there all the time*. We must therefore look far beyond drugs for the causes of drug abuse.

—2—
The Absentee Parent

Despite the strong sociological and peer influences at work on the adolescent, the matrix of the personality that turns towards drugs is developed *from birth* in the context of a *family*. Parents of a drug abuser are not necessarily "bad" parents. But the interpersonal relationships between parents and child are what form the pathological character necessary to create a drug abuser. The responsibility for the relationship, however, does fall heavily on the parents, by reason of age and maturity. As the child matures, a portion of that responsibility falls upon his own shoulders.

There are also innumerable familiar circumstances that affect life situations for the adolescent: death, divorce, siblings, and parental conflicts. How important are these situations in causing a child to abuse drugs? For a long time, the psychiatric professional has been aware of the familial and situational causes of psychopathology. One of the ways in which people handle anxiety and turmoil is by using drugs. Over the years, alcohol, opium, and cocaine have all been used for satisfying emotional needs. Drugs are tools of a defense mechanism used, as any other defense mechanism, to allay anxiety. This defense is more available and more appealing at the present time for a variety of social and cultural reasons, all of which tend to converge on the simple fact that today as never before parents are neglecting their responsibilities.

It is apparent to those who work closely with the drug-abusing adolescent that an "absentee" parent plays the major role in the specific psychopathology that leads a child to the use of drugs. There are many ways in which a parent can be absent from the home or the child. The most obvious occurs in situations of death and divorce. However, these are not at all the most prevalent ways in which a parent may be an absentee.

Most frequently, the absence that leads to drug abuse is emotional. The emotionally absent parent is one whose own needs supersede the needs of the children.

But let us first consider physical absence in some detail, since the mechanisms at play in such absence are models of those in emotional absence. Death is, of course, an unavoidable cause of absence, which affects the child in numerous ways. Some of the variables in this situation are: whether it is the mother or father who dies; the age of the child at the time of the death; the substitute relationships that the child has formed which can replace parental relationships; the presence of siblings, older or younger. The relationship with the deceased parent is of great importance.

Surprisingly, it is easier for a child to adjust to the loss of a parent with whom he or she has had a good relationship than one with whom he or she has had a poor one. Poor relationships with parents are always marked by ambivalence rather than by feelings of unconflicted animosity. Ambivalence is said to exist when a person loves and hates simultaneously. When a child feels ambivalence toward the dead parent, he has to cope with tremendous feelings of guilt because of the death. This guilt is stimulated by an unconscious death wish towards the parent, who at the same time he loves and needs. The death wish has to be repressed because it is too guilt-provoking to be handled comfortably by the child.

The guilt due to ambivalence is worse in early childhood than it is as the child matures. This is because there is a normal degree of ambivalence in all young children towards their parents. In addition, most children have a feeling of omnipotence, which is also normal in early childhood, and which diminishes as the person matures. The child feels that he or she is all powerful to the extent that his or her wishes can come true. This adds to the necessity of repressing other highly charged feelings of childhood that produce guilt.

When a young child loses a parent through death, he or she will frequently blame himself or herself as being the cause of the death. This blame is increased if there is an increased hostility towards the parent because of his or her heightened ambivalence. Even though feelings of omnipotence do diminish throughout life, the remnants always remain. Whenever a parent dies for whom the child possessed ambivalent feelings, traces of hate cause the child to question his or her responsibility.

When a child loses a parent to whom he or she feels a *minimal* amount of ambivalence, he suffers "grief." Grief, as understood here, is a normal feeling when one loses a loved one. It is generally self-limiting; after a period of mourning, one gradually experiences relief from the intense unhappiness and discomfort generated by the loss. When a child is highly ambivalent, however, rather than experiencing grief he or she becomes depressed. Depression, which is often confused with grief, is an abnormal reaction, which does not diminish with the period of mourning, but often becomes more intense. It is brought about by turning the anger and hate that one feels towards the loved one against oneself. It serves the dual purpose of expressing the feelings of anger which are demanding release, and punishing the self for the guilt growing from the belief that the anger is wrong or unpermissible. Depression plays such an important role in the drug abuser's personality that it will be discussed later in a separate chapter. It is heightened in the case of the death of a parent because the death wish cannot be justified in the child's mind.

The child's feelings of his own omnipotence are often mirrored by the feeling that the parent is also omnipotent. Young children expect their parents to perform any task that the chil-

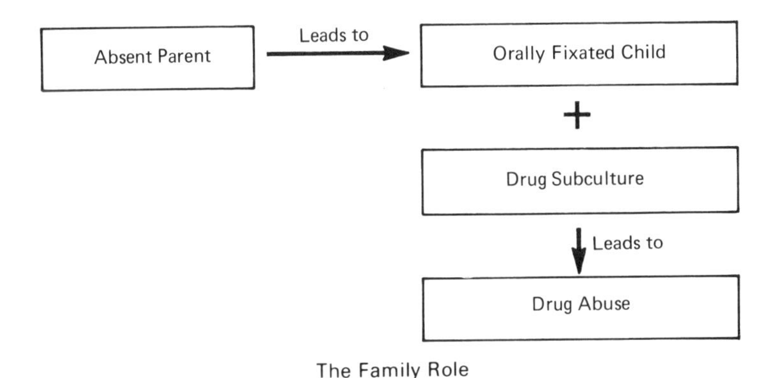

The Family Role

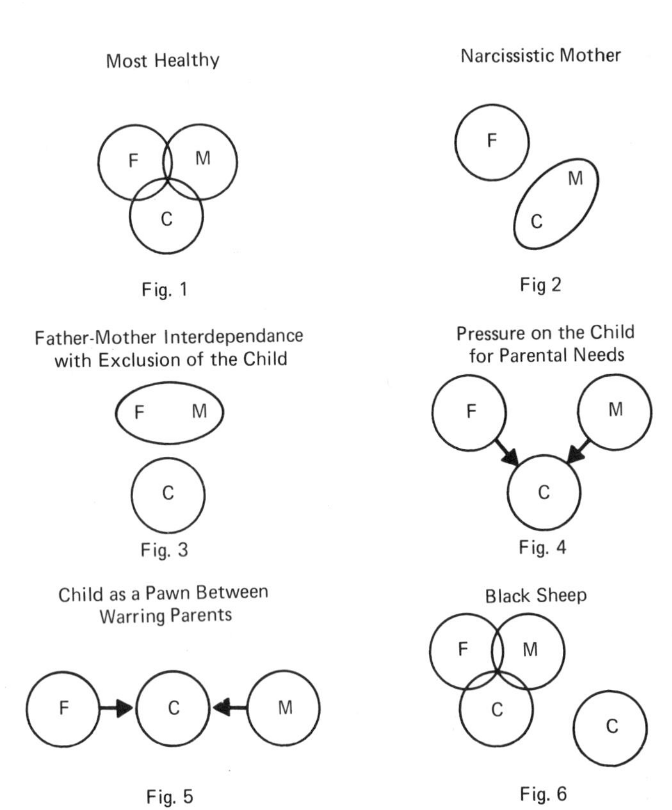

Various Family Constellations

Fig. 1 Each person maintains his individuality, but interacts to form a family unit.

Fig. 2 Mother and child become fused so the child can never become an individual.

Fig. 3 Mother and father are so dependent on one another that child is excluded.

Fig. 4 Father and/or mother, never being gratified by the self, turn to the child to provide them with gratification. (i.e., Child must become a sports star or an actress to satisfy parents needs.)

Fig. 5 When parents are in constant battle the child often becomes the pawn, often to instill guilt in the other parent, or for the parent to have an ally.

Fig. 6 When one child is special to the parents, the other may take the opposite tack, and cause further alienation.

dren feel they want done. A young boy in the midst of a fight with his friend will suddenly challenge, "My daddy can beat up your daddy." The injured child runs home to mother, whose kiss is far more healing than a doctor's medications. As healthy as it is, the belief of omnipotence in one's parents creates many mixed feelings when that parent is ill or dies. This belief creates an unconscious feeling that the parent could have prevented his own illness or death. Unwittingly, the child feels anger towards the parent for dying. In a child's fantasy life, the parent must be always present; so the deprivation of the parent in itself creates anger and hostility.

The reactions of a child to the loss of a parent are well-exemplified in the case of a young girl we shall call "E." E was sixteen when she was admitted to a psychiatric hospital. She had been probated there by the court, after having been arraigned for the selling of drugs, grand larceny, and car theft. She came from a white, middle-class background and was then living with her mother on the south shore of Long Island. She had gone through her junior year in high school when she began to be truant and then totally dropped out of school.

E's problems began when she was about thirteen years old. Superficially, they were behavioral problems, causing more consternation to her mother and the school than they did to E. Originally it was a case of truancy and bad conduct in classes, but she soon began to travel with a group of "tough" youngsters several years older than herself. At the age of fourteen she was involved with drugs to the point of selling them. At fifteen she was already classed as a juvenile delinquent—shoplifting, stealing cars, and committing minor burglaries. By the time she was sixteen she was totally uncontrollable and was well-known to the police in the area.

When E entered psychiatric treatment she was completely unmotivated, remaining in the hospital setting—and then in outpatient treatment—only because it was preferable to jail. As soon as probation was lifted, she dropped out of treatment and again found herself in difficulty. Despite all of the behavioral problems that she manifested, she had never "acted out" sexually. After leaving the hospital, however, she began to live with a heroin addict, who had an extensive police record himself. She turned to prostitution, briefly, in order to help her boyfriend support his habit. She began using heroin and mainlined on three or four occasions, but quickly returned to her favorite drug, "speed." She was caught smuggling dexadrines into jail when her boyfriend was arrested and once again she was placed on probation—into therapy. After six months of therapy she was still not motivated, and broke off treatment. Six months after that, however, when she was off probation, she once again returned to therapy for the final time, totally on her own initiative.

This far from unusual case history can be understood only against her parental background. E was born in New York but soon the family moved out of state. After two or three years they moved again and then finally to Long Island. E had always been close to her father during these early years, but she described her mother as being cold and aloof. She and her father would often go fishing together and enjoy one another's company in general. Because the father suffered frequent illnesses, he was home a great deal; her mother had to go out to work to help support the family. E remembers, when she was very young, often waiting behind the front door for her mother to come home—as she was constantly fearful that her mother would not return.

At the age of nine her favorite uncle, to whom she had been very close, died suddenly. Shortly thereafter her dog was killed. The final blow came at age eleven when her father died of his illness. At the time she was unable to mourn her father's death and remained stoical throughout the following few months. She claimed that she had to be strong in order to protect her mother, who was showing obvious grief over her loss. On reflection, E recalls that shortly before the father's death, she had begun to grow away from him. She no longer went on fishing trips and often refused to accompany him on other activities. She was not clear in her own mind why there was a sudden separation from her father. Though she had moved away emotionally from her father, she did not grow any closer to her mother and soon felt quite isolated. The feeling of isolation was complete following her father's death.

Throughout her early life E had few friends because of the family's numerous moves. She barely had a chance to establish roots in a given place before she was moving again. When she entered adolescent life, she still felt the absence of friendships. She had no boyfriends, and whatever heterosexual relationships she had were superficial and never progressed beyond petting. As she grew older she gravitated toward emotionally disturbed, weak boys whom she would tend to "put down" in an angry mood. She was unaware of this behavior until she reflected upon it during her therapy. Her relationship with men was always based on a sadistic need, in clinical terms, to "castrate" them.

Her early sexual memories include her fascination in watching her father get dressed and undressed, spying on him for a glimpse of his genitals. She would often sneak upstairs and peek into the partially open doorway as he was getting dressed. Later in adolescence she became quite an exhibitionist, often walking around nude even when there were strangers in the house. She engaged in homosexual activity and could not achieve orgasm during intercourse without cunnilingus.

During therapy it became apparent that E was wrestling with strong oedipal feelings that, because of her father's death, she could not resolve. The pushing away of her father in early adolescence—just before his death—was due to oedipal feelings she could no longer tolerate as her adolescent sexuality began to develop. She needed to deny the desire for her father. When he died, therefore, she was unable to face the fact that she felt a loss, so she was unable to grieve. Yet, she did feel an intense loss unconsciously, with a resultant anger which she could not justify and which brought about guilt.

This chain reaction continued as, in order to repress guilt and anger, she began to deny a need for any family member. This denial required pushing away all family values and mores and acting out the anger in an antisocial manner. She remained without grief and without feelings, taking out her anger in sadistic relationships with men until she was well along in therapy.

Through the transference mechanism, she began to become aware of a strong dependency need for the therapist. The therapist interpreted to her that this was really a need for her parents, a need which she had previously denied. Once this was recognized she entered a belated period of mourning over her father's death and returned to a relationship with her mother. Recognizing that her absent mother had caused her to be quite unhappy during childhood, she established a new relationship with her, based upon the needs that she had as a child. She now recognized that she had the need to repress the desire for closeness to her mother because her mother was absent not by choice but of necessity.

E illustrates the intense rage in a "secondary" acting out that a child feels at the loss of a parent. (It is "secondary" in that it goes beyond a simple mental vision of a need: her attitude toward school, toward men, toward courts, and finally toward the therapist.) Despite an overt denial of parental needs and a rebellion that became so intense as to take on a totally opposite value system, E was really suffering from intense *dependency* needs, especially towards her father. The underlying depression over the loss of her father, her dog, and her uncle could not be articulated because of her complex desire to avoid the recognition of her parental relationships. It is just such rage and subsequent depression that so frequently turns a child to "acting out" behavior. Acting out helped repress E's pain of depression. The only reason she came into therapy originally was to satisfy the court.

Serious illness requiring hospitalization or confinement to bed over prolonged periods has an effect similar to the death of a parent. Because the parent is still alive, however, it is often more confusing for the child who has to confront his ambivalence person to person. The anger towards a sick parent is acted on in many ways, but not in a conscious manner. As in death, a loss is felt, but, as the burden of responsibility falls *to the child* in replacing the sick parent, resentment continues to grow. This resentment, by necessity, must be repressed because the child cannot justify anger at a parent who is sick and *because* he is sick. The child has learned at an early age that one must pity any ill person and not feel angry towards him.

The case of C, who started using barbiturates at the age of thirteen, illustrates this point. From the time she was eleven her mother had had frequent illnesses requiring hospitalization. Prior to her mother's first illness C had experienced many angry outbursts from both of her parents. In her family, to be angry meant to be violent; physical beatings were suffered at the hands of both parents when they became incensed at her. She and her brothers and sisters would be beaten with coat hangers, belts, or any other convenient weapon that their parents could lay their hands on. Even when the parents fought only with themselves, physical abuse was the rule.

Whenever her mother became ill and required the hospitalization, C was left in charge of her brothers and sisters because she was the oldest. Her father became very demanding of her. She began to feel that he believed she had to replace his wife. He did not make allowances for C's age and C felt constantly guilty if she was unable to maintain her mother's standards in running the home.

It had always been unacceptable for her to be angry with her mother for being sick. She said to herself, "She is so good to me usually, how can I *not* do things for her when she is sick?" Through therapy, however, it began to become clear to C that she herself wanted the mothering. She became frustrated and angry when she was the "mother" to her three younger brothers and sisters. Though it produced a great deal of guilt, she would often release this anger toward her siblings. She would often spank them at the slightest provocation. Her infantile need had always been frustrated whenever her mother became sick, but as she grew into adolescence she could not accept that she still had these desires. So she repressed her anger and developed the subsequent depression which the barbiturates were taken to alleviate. Secondarily, the barbiturates chemically relieved both the guilt which she felt because of the anger, *and* the anger.

Unconsciously, the taking of drugs was a device to cause a restructuring of her family. It was not long after she began using barbiturates that her parents became aware of it. She often purposely came into the house in an intoxicated condition. It was only then that her parents reassumed the parental role which she so badly sought but could not admit to herself.

The case of C illustrates how the loss of a mother through illness exhibits the same mechanism as a loss through death, causing even more confused feelings in the child. The circumstances in the cases of both E and C were unavoidable. The death and the illness could not have been circumvented; nonetheless, the infantile needs for the parents had not been met.

There were no suitable substitutes for the loss. E could not turn to her mother for comfort as she was too cold and remote. C could not turn to her father as he had expected wifely duties from her and had stopped treating her like a daughter. This inability to turn to another parent figure increased the anger towards the deceased or missing parent and created severe resentment of the remaining parent.

Anger towards a remaining parent is at least as uncomfortable as anger towards a deceased or missing parent. The fear of the loss of the only one to whom the child can turn or the fear of the loss of love becomes so intense that no negative feelings can be permitted towards the remaining parent. Without the relief of being permitted a negative feeling, only repressed anger can result. The anger that is felt towards the deceased or towards the remaining parent is not, of course, a mature, rational anger. As an infantile form of anger based upon self-centered demands for the parent, it has its beginnings in infancy when such demands are normal. The *unjustifiable* nature of the anger causes it to be unacceptable and therefore expressed as depression.

Parental absence by divorce is a more common phenomenon in the modern family than absence by death. The likelihood of a marriage ending in divorce is approaching one out of three, but of course it is considerably less where children are involved. The instability of the social structure is mirrored in the marital structure, with an apparent decline in commitments in all areas of life. People move their homes frequently; few families have roots in one area for more than a generation. The younger generation changes jobs and colleges as readily as marital partners.

As loyalty to any of these once "sacred" institutions evaporates, parents wonder why their children lack commitment, goals, and direction. It seems rare to find an adolescent today with a set goal and the will to pursue it. Apparently, the couple with the goal of maintaining a marriage and with the commitment to solving problems rather than separating is also a vanishing breed.

Divorce presents special problems for the child. Once again he is confronted with a parent who is absent, yet in a more threatening way. While the death of a parent brings about a final parting and intense grief, divorce involves no such finality. There are two situations vis-à-vis a child in a divorce: one in which the absent parent has visitation rights and sees the child on a regular basis; the other in which the parent totally absents himself from the family. Each presents a special problem to the child. In the former situation, a child has to contend with the repeated coming and going of a loved one. This causes a repetitive resurgence of anger and of resulting depression. The child has to contend weekly with the mixed feelings involving the parent who is gone. He is glad to see the parent, but at the same time has to suppress the anger caused by the parent no longer living with him and therefore no longer available to meet his needs. The child begins to ask himself why the parent has left. Is it his (the child's) fault? The fantasy is often that if the parent had loved him enough he would have stayed. The child is well aware that it is a matter of choice in the parent's leaving and therefore differentiates it from the parent's dying. The unconscious infantile fantasy that the parent could have remained alive is counteracted by the conscious knowledge that he did not choose to die. In a case of divorce the child is fully aware that the parent could have remained if he had so chosen.

The trauma of a divorce situation starts long before the actual separation. Many parents like to delude themselves into thinking that the children are not aware of the strife and discomfort that is taking place between them. This is very rarely the case. A child is highly attuned to the parent's moods, feelings, and arguments that occur even in the most discreet situations. No matter how enlightened the parent may be in trying to keep the child free from the anxieties of the situation, the child is already developing feelings of depression and anger prior to the divorce. By necessity, most of these feelings are being repressed because they are too uncomfortable for a youngster to handle.

Divorce of parents is more traumatic for younger children than it is for adolescents. The adolescent finds it easier to withstand the tension of the divorce situation on a rational level. In the younger child, fear is the most common emotional experience. The fear of loss of the parent is paramount, but even prior to the actual loss there is an overwhelming fear of the loss of *control* that the parents often express during their fights. The young child who looks to his parents as a role model idealizes them and imagines that they are capable of solving the child's problems. The witnessing of the loss of control by the parent, such as in a fight or in a situation of rage, when he storms out of the house, becomes very confusing to the young child. Whom does he turn to at this time? Where does he look to find stability and security? Where are the people who have all the answers to his problems when they cannot handle their own?

These questions constantly plague the child as he begins to withdraw into himself. He learns to distrust the environment and the people in it. He knows that he is not capable of sustaining himself and yet he begins to turn to himself for answers. There is an air of independence surrounding the child who feels that he has no one but himself; but behind this facade of independence is fear, lack of self-confidence, and a strong need to attach himself to someone or something. Because of his distrust, he does not allow anybody to become close enough for him to turn to. He becomes lonely and more frightened and, as he does, the "pseudo-independent-dependent character" develops. Unconsciously, the child is aware of this façade, but nonetheless it has to be maintained. Unfortunately, throughout his life, he never learns that he can drop the wall that he has placed between himself and the world and it becomes difficult for him to relate to others or to relate comfortably to himself.

Even in the instances where there is no overt hostility between the parents, there is a tension that affects the child. The subliminal feeling of instability is recognized. It is usually unspoken by the parents and unexpressed by the child overtly, but the child becomes moody, morose, and withdrawn, or he may exhibit a false bravado as if no problem is present. His superficial appearance, however, does not reflect his internal turmoil. An apathetic attitude develops towards his parents and the parental values. He develops this apathy in order to deny to himself the importance of the love he has for his parents. Actually, it is an attempt to diminish the anticipated hurt, "to quit before he gets fired" by his parents. This show of independence gives the child a feeling of control over the situation which he fears and lessens the feeling of helplessness.

Much of the child's unconscious rebellion takes place prior to the parent's leaving. The anticipation of the event is often worse than the reality of the loss. Protective mechanisms begin to develop prior to the actual separation. When the separation finally takes place the child responds internally with confused and unacceptable feelings. The younger the child the less he is able to understand his feelings of anger and guilt. The feelings of abandonment are experienced as a punishment for a vague crime, causing unrealistic guilt within the young child as he blames himself for his parent's leaving.

As mentioned previously, the young child is still wrestling with feelings of omnipotence, which lead him to believe that he was responsible for forcing the parent away from the house. What he relates as he gets older is, "Maybe I could have done something to prevent the divorce," or "If he loved me enough he would not have left; maybe I did something to make him stop loving me." The anger at a parent for leaving is not easy for a child to understand and to accept. Even when the child loved the parent, and the parent and the child had a good relationship prior to the divorce, much of the anger has to be repressed. The repression is necessary for fear of further loss of the parent. The child becomes afraid that he will lose what little relationship still remains between him and the absent parent and that his anger will hasten this breakdown of the relationship. One of the mechanisms used to repress the anger is to turn it against himself—and with this the child becomes self-destructive.

In our society it is more frequent for the child of a divorce to remain with the mother; the loss of the father, accordingly, is most commonly experienced. In the preadolescent child there is a strong dependency attachment towards the mother for nurturing. The fear of losing the mother, therefore, is greater than the fear of losing the father. This is true for children of both sexes and is independent of oedipal feelings. It is based primarily upon the oral needs of the individual for a nurturing figure. The mother is seen as the source for nurturing and for love, and a child is happy at having this all for himself. He does not want to share it either with siblings or with the other parent.

It is often the fear of his father's anger that keeps the child from pushing him away. Nevertheless, we often hear a young child say, "Go away, Daddy," as he turns his back on his father and clings to his mother. This clinging will diminish, it is hoped, as the child matures. But in boys the attachment remains strong. With the onset of the oedipal stage, girls begin to turn their attachments to their fathers, while boys remain with the mother. The most distressing part of this emotional conflict is the unconscious desire of the child to rid himself of the parent of the same sex, so the parent of the opposite sex could be possessed solely.

In the normal course of events the oedipal stage is resolved between the ages of eight and ten—accompanied by the child's identification with the rival parent. In the process of identification, the child will take on the characteristics he most admires in the parent and will turn his amorous feeling towards someone outside the family with characteristics often like the loved parent. In a divorce this situation is difficult. The realization of the unconscious wish to be rid of the father by the young child leaves him with excessive guilt because he blames himself for the father's leaving. The son maintains these feelings throughout the oedipal stage, but the daughter now has entirely new feelings with which to cope.

Prior to age four, both sons and daughters share the similar feelings of the desire to be nurtured, but after the onset of the oedipal stage, even though these feelings change, there are always remnants of pre-oedipal, dependent feelings that linger within the girl. This tends to confuse her further; part of her wants her mother for nurturing while part of her wants her father for loving. She begins to wonder if this infantile sexual feeling towards her father did something to drive him off (to divorce). The situation is repeated in adolescence when the girl finds it difficult to form a comfortable sexual relationship with a man, being fearful that the sexual feelings of infancy which, in her mind, "caused" her father's leaving will repeat themselves and cause the man she loves to leave her also. In the adolescent girl these feelings can lead either to frigidity due to repression of all sexual feelings or to promiscuity without a love relationship. Sexual promiscuity is further promoted by the desire of the girl to recapture her father.

At this stage promiscuity takes the form of an *incorporation* into herself of a male figure, through the male organ. There is usually no orgasmic sexual pleasure from this relationship and the greatest pleasure is derived following the sexual act when the couple lies together in a nonsexual union. Guilt over the sexual act is intense, as sexual feelings are still being blamed for the loss of her father. This guilt is complicated further by the cultural and family teachings of the wrongness of the sex act.

Without a man in the house, it is difficult for a son to resolve his oedipal feelings towards his mother. In essence, his unconscious oedipal wish has been realized and he is now the man of the house. Yet when a son loses a father, he has lost his primary means of identification. The process of identification is a necessary part of the resolution of the oedipal complex. Without an object with whom to identify, the son flounders. The guilt at the possession of the mother will often drive the son away from the mother and lead to constant conflict between them. Much of the fighting that is seen between a son and a mother in a divorced family comes from the very need of the son to push away the object of his sexual desires which, because of its unacceptability, causes anxiety. Any affection between the two is confused as sexuality and therefore is unacceptable to the son. Through it all, the intense resentment and anger felt towards the father for the desertion remains.

Despite the oedipal desires of ridding himself of the father, a boy's identification with his father begins at around age two. It is the ambivalence towards the father which is characteristic of that age, and not outright hate, that is the hallmark of the oedipal relationship. When the "love object" has abandoned the son, therefore, the son becomes mistrustful of men and desires to cling even more strongly to his mother, throwing him into an even greater turmoil because of increasing oedipal feelings. He has attempted with all his strength to form a relationship with his father, and if his father is available to him, at times there will be a partial identification. If the father remains aloof and physically away from the child, however, then there is no object for identification and the son will tend to identify with his mother.

Strong homosexual feelings often develop in this situation. First, the son has no male identity. He is being driven to identify with a woman towards whom he has taboo sexual feelings; therefore, his heterosexual feelings have to be repressed. Furthermore, as he grows into adolescence, the desire of a male for identification becomes confused with the desire of the male for sexual reasons. The homosexuality is rarely overt. Nonetheless it is very frightening to the growing adolescent to have to contend with his homosexual feelings. The feelings themselves often cause a child to seek relief in drugs to further repress homosexuality and alleviate guilt.

It is not uncommon for adolescent boys under the influence of drugs, especially the barbiturates, to indulge in homosexual activity. This behavior is repugnant and guilt-laden to them, and would not be indulged in when they are straight. The rationale the adolescent uses is that it was caused by the drug and not by his own desires. This rationalization allows him to accept himself in a situation highly alien to his ego.

The loss of a mother due to divorce is a more traumatic situation than the loss of a father. The father has been absent from home on a daily basis when he is at work, while the mother is primarily the person who spends most of the time with the child. It is difficult to obtain a large sampling of youngsters whose mothers have left in a normal situation. Until recently, only when the mother shows signs of severe emotional illness or undue cruelty to the child will the court give custody of the child to the father. In the following cases, it must be understood that the mother showed a severe degree of emotional pathology, often diagnosed as schizophrenia, with numerous hospitalizations.

It is not difficult to imagine what it is like for a small child to lose his or her mother. Any parent who has left a child with a stranger, even for brief periods, recognizes the changes in the child upon his return. Because of our sociological structure, a motherless child is almost always a fatherless child. He or she is usually kept with a housekeeper while the father is off at work. In contrast, a divorced mother receives alimony and child support from the working father and can often spend most of the day at home. The child in custody of the father feels abandoned and alone; to a young child this is terrifying. No matter how strong an attachment the child develops to the housekeeper or babysitter who takes the mothering role, he or she is aware that this is not the mother. In the preoedipal child or even in the child with oedipal feelings, the dependency need is intense.

It is not within the scope of this book to discuss all of the possibilities that develop in a motherless home; we must limit the discussion to the drug-abusing personality. Most commonly seen in the victims of drug abuse is a repression of dependency needs and a development of pseudo-independence. There is constant distrust of close relationships, as the primary relationship has been broken. The constant fear that there will be a repetition of this loss plagues the individual throughout life, and he finds himself unable to form a love relationship. A depression ensues with the loss of a loved one, and the depression must be repressed along with the dependency need, because to admit to the depression is tantamount to admitting to the dependency. "I don't care," is the apathetic statement that one most frequently hears. "It didn't matter to me that she left because I didn't need her anyway," is expressed in one form or another, both in words and in action. As with anger at the father, the anger at abandonment has to be repressed because of the fear that the remaining part of the relationship will be destroyed because of the child's rageful feelings. The rage is then often projected onto the world, making the child even more distrustful of human relationships.

As the abandoned child grows into adolescence, a boy replays the ambivalence towards his mother with the women in his life. There is a strong attachment to girls and yet equally strong feelings of rage. He is fearful of the intensity of his feelings and tries to minimize the importance of the relationships to himself. Such fears lead to grouping, to not "putting all one's eggs in one basket" when it comes to heterosexual relationships. Because the boy feels that he is attached to the girl and that he has conquered her at the same time, his relationship is sexual and not one of love. Often the boy will avoid girls entirely in order to "play it safe" and not have to contend with his intense but dangerous desires.

Girls will often continue to seek a mother figure, either in terms of a group of girls or in terms of a man. Here again, sexual promiscuity is often witnessed as a girl seeks a mothering figure in the male. The penile intrusion with the subsequent male orgasm has been likened to fulfillment of feeding in early infancy. Once again there is no orgastic pleasure for the girl from the sexual act; rather, she merely desires the fulfilling feeling of the penis. One young girl said, "Sex would be so enjoyable if the boy didn't move around so much and just was willing to lie there."

There is little desire for this type of person to *give* to others, but a constant need to *receive*, as in infancy. Subsequently, there is a consistent anger at all the people to whom attachments form for not giving enough. This is true of both boys and girls in a motherless situation. The anger caused by the absence of the mother is then expressed in all the peer relationships that they develop throughout adolescence. They want to be "fed" with love and concern and at the same time not to be required to give anything in return. All subsequent relationships that develop seem to be based on this infantile one. There are few meaningful adult relationships throughout life for them, and the chances are that they will perpetuate in their children the same difficulties that their parents had bequeathed to them. As it is difficult for them to form a meaningful love relationship and there is resentment at not having the dependency need constantly fed by the marital partner, the second generation marriage frequently ends in divorce.

The assumption that an adolescent can withstand the rigors of a divorce on his or her own is contrary to the evidence. Many couples who have lived with strife throughout their children's younger years decide to divorce once the children enter their teens because they believe that the children are better able to withstand the trauma. To an extent this is true, as the children's intelligence and judgment enable them to understand the situation that is taking place. Yet along with the understanding comes the need to repress the feelings about the divorce. An adolescent, being in the throes of developing his own independent life, by necessity will deny and repress any dependency needs.

It is for this reason that the normal adolescent appears to be in such turmoil and becomes so rebellious and irascible. This is a normal growth process of adolescence. In the situation where dependency becomes threatening, however, because of the fear of loss of the person upon whom he is dependent, as well as threatening because of fear of his own internal dependency needs, then the denial and repression become seriously intensified. Until the early twenties the child is still feeling the conflict between dependency and independency.

When the parents of a child of this age divorce, he is placed in the situation of feeling the need for the parent who has left while simultaneously having to *deny* the need for the parent who has left. He becomes morose and distant. The emotionally absent parent may observe the same reactions in his children and ask himself, "If I am not dead or divorced, what am I doing here?"

—3—
The Distant Parent

We have seen how a dependent youngster, thrust from the security of a strong family nucleus, finds it necessary to form his relationships elsewhere. He typically bases them on the early association with his parents. If an adolescent does not feel the comfort of such familial relationships, all subsequent relationships are going to be superficial and fraught with anxiety. He is a perfect target for the drug-abusing crowd, for the superficial relationships based on drug use rather than on his peers.

In these situations especially, the neurotic relationship of the parents toward their children leads to a tendency towards the drug structure. When an ambivalent love-hate interplay exists between parent and child, the fear of forming close bonds thrusts the child into superficial friendships. Adolescents find themselves in mutually dependent relationships. As sexual taboos are now all but gone, a youngster can seek a sexual or communal existence without entering the prolonged orientation that precedes a close relationship in a normal society. The root of the problem, again, is the "absenteeism" of the parent.

Yet far more common than a physically absent parent is an emotionally absent parent. We might call him the distant parent. Either the parent is so narcissistically self-invested that his needs supersede those of the child, or the parent's involvement with a career becomes paramount in his existence—child or no. Narcissism seems to be found predominantly in a mother, while excessive career involvement predominates with the father.

The mother can manifest her narcissism in numerous ways. It can either be an overt manifestation of her personality, or it can be more covert and therefore harder to pinpoint. An overtly narcissistic woman obviously is the one who is openly more involved with her own needs than with the needs of her child. She is conspicuously absent from her child's life during the formative years. She may be involved with a career or with clubs and organizations or volunteer work. Not infrequently she is intently involved with organizations that are child oriented, such as the P.T.A. and other school activities. This involvement seems to compensate for the guilt caused by her absence from the child; she sees her involvement as being for the benefit of the child. The woman's own aggrandizement, however, predominates; she feels she has a legitimate excuse to escape from the responsibilities that she should have toward her child.

It is often difficult for a modern woman to reconcile herself to the sacrifices that are necessary to be a mother. In increasing numbers, she has been highly educated and highly trained prior to marriage and to motherhood, and she is suddenly asked to put all of this aside for the sake of raising a child. One of two situations may arise. First, she may return, almost immediately, to the career embarked upon prior to motherhood, leaving the child in the care of a housekeeper or permanent baby sitter. Second, she may remain at home with the child, with such severe unconscious resentment of the interference with her desired life style that subtle hostility is rained upon the infant. In the former case, the child responds as to any absent parent figure as we have seen. In the latter, a hostile relationship that develops between the mother and child can be perpetuated throughout the life of the child and permeates all of the child's later relationships.

How is this hostility meted out upon the child? It can either be through direct attack or through indirect attack. There is often an overreaction to the child's naughty behavior, where punishment for his actions grows far in excess of what is necessary. Children have been subjected to intense beatings and abuse over the most trivial matters. The mother rationalizes that she is merely disciplining her child in the normal process of child rearing. These punishments cause an often unspoken reaction in the child, making him wonder what he has done to deserve such treatment. The child is unwilling to blame the parent for being at fault and takes the fault upon his own shoulders. He develops a profound sense of guilt and undesirability. He assumes that the parent is justified in that form of punishment and that he must have done something of which he is unaware to have stimulated such abuse. Any overt reaction by the parent seems only to confirm the child's conviction.

Many verbal barbs are directed by the mother towards her child. As many of these children have unconsciously been unwanted, a mother may communicate this additional fact to the child. A number of youngsters have related how their mothers told them that they were conceived by accident, that they were unplanned. They make the only assumption possible: they are unwanted. The mother justifies this explanation as being honest with the child; the child "will feel an increased amount of love knowing that he is a 'love child.'" The child consistently feels, however, that he is an unwanted burden upon the family. He will often attempt to please the family to such a degree that he gives up all individuality, never being able to feel that he pleases his mother adequately, or he becomes sullen and hostile and will often take the opposite direction. He may also turn against the family, becoming rebellious, seeking an unacceptable group of friends. They turn from school and again become easy prey to the drug culture.

In essence, the child is correct in his assumption. With a narcissistic mother, no matter what the child does as an individual, he cannot possibly please her. The narcissist demands that the child's needs, goals and direction parallel her own. She does not allow the child to deviate from her outlook and to become an individual without marked resentment and hostility on her part. There are more than the usual fights and arguments whenever the child tries to make a break from the family style. The parent justifies this to herself by claiming that she knows what is best for the child; in reality, the resentment is that the child is not fitting the exact mold that the mother demands.

In the child's struggle for independence, this repression takes on serious consequences. He becomes threatened with any step out of line with the family thinking. He feels that any independent action will be displeasing towards his parents and therefore dangerous. He often reacts by following his parents' footsteps almost exactly.

Such a child never goes through normal adolescent rebellion; rather he is always considered a "good" child. Unfortunately, when the time comes to make the break, in the late teens or early twenties, the child does not feel prepared to assert his own independent action. He constantly feels the need for the parent to be looking over his shoulder to guide and direct his life. However, the realization *that independence is now required* becomes threatening. It is then that the child will generally make a break with an explosion that is loud enough to rock the entire family.

As frequently happens, children of this type are away from home for the first time to attend college, or they move out of their house and maintain their own apartment. The more tightly bound they feel towards their parents the more intense the explosion to make the break. As they do not feel confident to turn from their established path in a productive direction, they usually take the path of least resistance. To become productive is exceedingly threatening because they do not feel that they can succeed in a manner which differs at all from their parental guidelines. The course of least resistance, therefore, is totally nonproductive. They feel that school or work is posing too much of a threat to their own integrity. They will drop out of school or quit their jobs and drift aimlessly for a number of years. Consequently, parents go into turmoil over their "good" children. Because the children experience a secondary gain of causing anxiety and aggravation for their parents—for whom they have unconsciously harbored resentment for years—it is that much more difficult to retrieve them from their drifting.

All children have the desire to be nurtured and to be allowed to grow and develop on their own. It is at the age of two that a child first makes an attempt to assert his own independence from his parents. The child begins to recognize the unwillingness of parents to allow him to fulfill his own needs. Unconscious resentment begins to develop at that time and continues throughout life; it is highly intensified in adolescence, when independence becomes necessary.

Conversely, a mother who is not attuned to a child's needs may become overly neglectful and therefore fail to set limits at all, allowing her child to grow in what appears to be an independent situation. The term "permissive environment" has often been used by a neglectful parent to condone the neglect. In the normal course of development, a child needs limits that are set consistently and firmly, along with his freedom to grow.

As with most things in life, the middle road in child rearing is the one that raises the healthiest children. By not setting any limits in a child's life, parents find the child developing fears of his own unconscious drives without the ego strength necessary to keep these drives in check. He will always be looking for limit-setting authority, which often necessitates his acting in an antisocial way: looking for the police or other members of an authoritarian group. He is crying out for what was lacking in his childhood.

It is clear that the neglect of a child's needs can fall on either extreme of the disciplinary range. Parents have often been accused of being either too lenient or too strict. *It is generally not the leniency or the strictness of the parent that produces a disturbed child, who can become easily enticed by the drug world, but rather it is a neglect of consistency and of a rationale behind the leniency or strictness.*

Authoritarian parents, who set very strict limitations on their children, with their needs kept in mind, can produce healthy children as well as permissive parents who allow their children a great deal of freedom. The freedom in this case is again within limits. It is unreasonable, however, to believe that a child who is treated differently from the majority of his peers is really being considered in terms of his own needs.

In the adolescent years, when peer relationships take on such importance, equal treatment by parents becomes a necessity. Frequently a parent is so threatened by the child's breaking away from his set ways that he will unwillingly divert the child from healthy, constructive friendships to those that are destructive and hurtful to the child. A fourteen-year-old girl, for example, who is unable to go to the basketball game at the school gym with her friends because it is past her usual curfew, will very often turn completely around and become totally rebellious against her parental standards, running away and joining other overly rebellious adolescents. Healthy, constructive friends will no longer accept her because of her constant pulling away from social values. The only peers left to her are those who essentialy are in the same emotional bind.

When parents are out of tune with today's standards and today's adolescent needs, they do not let the child exist as an individual. It is more comfortable for them to force the child into their own mold, as this withdrawal diminishes their own anxiety. It is easy to understand how it would be more comfortable to have a child in bed every night by eight o'clock and not run the risks that late hours may incur. *The fear of letting go* and the preserving of the parents' own emotional comfort cause the child to feel neglected.

On the other hand, it is often easier for a parent not to set limits at all, to avoid all fights and confrontations that by necessity arise when there is a disagreement of desires. Some parents actually find it easier to allow the child to express his full range of desires. So they avoid the frequent arguments which arise when the child desires to go beyond the limits they have set.

Arguments and confrontations are necessary in the rearing of any adolescent; an adolescent is constantly straining at the leash to break loose. His disagreeableness should not be mistaken for blind rebelliousness. Permissiveness in child rearing, from its heyday in the thirties, has been based on the idea that the child should be allowed to grow and expand and develop far beyond the rigid limitations that are imposed by overly austere child rearing methods. Yet permissiveness and neglect are not synonomous. It is more difficult, actually, to be a permissive parent than to be an authoritarian parent because of the constant decision-making that arises when each new situation comes about. To live by a code of set rules that do not vary and do not bend to different situations takes the weight of decisions off the parents' shoulders. When one has to consider each individual situation in light of the circumstances that are involved, it is more anxiety-provoking for the parents. It is to be expected, therefore, that a parent who is basically interested in his own needs, above and beyond his child's needs, would not want to be bothered by this constant agitation.

The resentment of a narcissistic mother toward her child can make her appear overly protective. The overprotective mother, who does not let her child out of her sight for fear that he will be hurt, is usually responding to the internal, unconscious rage at the child's presence. Her overreaction takes the form of an overwhelming desire to protect him from the world.

One mother in our experience would not let her child out of his own back yard until he was thirteen years old. He was restricted from friends and activities because they were deemed too dangerous. This child, once he did leave the confines of his mother's home, was unable to form any normal relationships and found the anxiety of freedom too great to handle. He fell easily into a drug pattern which placed no demands on him and relieved him of the anxieties of an otherwise threatening environment.

Another mother made all the decisions for her son up until the time he was eighteen years old. She felt that because she "knew best" that she would protect her son from all the harm that might befall him if he were left to form his own judgments. The child never learned to deal with the world, of course, but learned to rely upon his mother for all his needs and decisions. He too became frightened when he was confronted with the situation of making his own decisions and of coping in a basically threatening environment.

It is clear that it was not for the child's needs that these mothers were overprotective, but rather to allay their own fears and anxieties growing out of their relationships with their children. In general, these fears and anxieties are developed by the parents' own unconscious rage at their children for interfering with their way of life.

Another example of subtle, narcissistic needs interfering with a child's upbringing is the use of the child as a parent substitute. This is often seen in instances where a mother is too lazy or too ill to care for the house. It was exemplified in the case we have already discussed of a child who, because of the illness of her mother, was forced into the maternal role. It became apparent, however, when she was taken from her home to enter the day care center, that it was not merely her mother's illness that required her to take care of the younger children and to run the home. Her mother had used her as a mother substitute, a reversal of roles. She had relied upon her daughter to care for the younger children and to run the home to be relieved of the anxiety caused by her own inability to do so. When her daughter was removed from the house, the mother became highly anxious and visibly agitated at the loss. At one point she threatened to remove her daughter from the center, even though she was obviously happy there. Surprisingly, this is not an unusual situation. Mothers will often use their children to handle many of the chores that they find too odious. The important point here is that these are not just the normal chores of childhood that take place in all homes. Rather, the child is relied upon to a degree that makes him or her the parent and the parent the child. The child cannot tolerate this responsibility as his own dependency needs are going unfulfilled. The resentment towards the parent is intense; the fear of his own incapability and consequently the fear of not being able to assume these responsibilities keeps the child in a state of constant anxiety, usually leading to the pseudo-independent façade.

In some instances the replacement of the spouse is even more overtly pathological. Commonly, a mother will sleep with her young son when the father is out of town. She rationalizes that this is because the child is afraid, but generally it is because of the mother's own loneliness and fear of sleeping alone. This practice stimulates highly erotic feelings in the child and in subsequent years causes unresolved anxieties.

More frequent than a mother-son relationship is father-daughter behavior that is overtly incestuous. The daughter is caressed and fondled by her father under the guise of affection. There may have been genital contact or stimulation by the father, and not infrequently the daughter has been asked to masturbate the father. Numerous youngsters recall episodes of having a father rub his genitals against the buttocks or even against the vagina during their early years. One young woman was a frequent witness to her father's masturbatory activities and was consequently unable ever to achieve full orgasm without fantasizing an old man in the act of masturbation. Needless to say, this type of overtly bizarre parental activity demonstrates a far greater degree of pathology than is usualy encountered. Nevertheless, it too is a symptom of extreme narcissistic involvement of the parent with his or her own needs, rather than consideration of the needs and desires of his or her daughter.

Since the years of women's suffrage and more recently of women's liberation, more and more women have become career and goal oriented. It is no longer deemed acceptable for a woman to be a housewife and mother. Rather, women say they do not feel fulfilled unless they are involved in an occupation that years ago was reserved only for men. There is certainly nothing wrong with this goal orientation; however, it seems that many women have avoided the responsibility of making the decision between career orientation and motherhood. They want both and are unwilling to give up either. They have rationalized a situation of role reversal, in which the father stays home and cares for the child while they go off to work.

Only rarely has such a sudden reversal worked well, and in many cases it develops a severe pathology in the individual. Many daughters have expressed resentment at the mother for not assuming the motherly role and turning it over to the father. Sons have expressed the same resentment and even a greater degree of hostility towards their father, who is the person with whom they want to identify and who is not asserting his "manhood."

Our concern here is not to make a moralistic judgment, but rather to identify a pragmatic situation that has occurred with many youngsters who have come from families where the parental roles have been reversed. There appears to be a higher degree of drug abuse in the family of an overdomineering woman and an overpassive man. This pattern also seems to give rise to a greater degree of schizophrenia within the family structure. Perhaps this is further evidence that the maternal instinct is not a myth, but rather a natural tendency that is shared by many women and by all children. Prior to the advent of bottle feeding, it would have been very difficult for a father to stay home and care for a nursing infant. The artificial implement of the baby bottle perhaps has led to artificial child rearing procedures.

With the increased number of women obtaining higher education and careers, there is an increase in general resentment at being "saddled" at home with a child. The alternative to such resentment is to leave the child and to pursue a career. Many women have returned to their careers shortly after the child is born, sometimes even within the first six weeks—leaving the child in the custody of a nurse or baby sitter. This arrangement becomes confusing to the child, as he loses the consistency of the object relationship. He does not have an opportunity in the early months of life to form an association with one maternal figure. The identity with the mother is split, and therefore the ego is split between the mother and the caring person.

Following the death of a mother, there is usually only one caring object in the child's life, whether it be a grandparent or housekeeper or stepmother. In the case of the career woman, the child is handed back and forth between the baby sitter and the mother when she returns home from work, with far greater confusion than if there were a constant object. Even in those situations where object-consistency is already established before the mother goes to work, the child soon feels confusion about the absence of the mother. He begins to wonder, as in the case of divorce, whether or not the mother actually cares for him as much as he wants to be cared for. An intense sense of competition and rivalry develops as the child senses, very intently, in very early years, the tearing of the mother between the career and the child. Such competition is not dissimilar to intense sibling rivalry, except that it tears the mother away from the child for long periods at a time. A resulting sense of abandonment continues even into adolescence.

A fifteen-year-old girl recently related to us how she felt intense resentment when the mother resumed her career the year before. She felt that she was returning from school to an empty house and to no one with whom she could talk about her daily activities. She had looked forward with great anticipation to the daily chats with her mother, and up until this time had no intention of drifting from her family life. As the child was no longer returning to a warm, accepting household, she began to drift into the street after school and, finding that she was not receiving maternal gratification from her usual peer relationships, she tended towards those based more upon mutual need.

It must be remembered that a drug relationship is not mature but rather based on mutual interdependence. The youngster feels that the other chlid needs him or her as much or vice-versa, and therfore is safe. This mutual interdependency forms intense infantile bonds that are difficult to break. In a healthy peer relationship, in which a very dependent child forms a relationship with an independent peer, that peer does not have to be available to the dependent child at all times. He will have other friends and other interests that are exclusive of the dependent adolescent's needs. In a mutually interdependent relationship, however, one will not function without the other. As the relationship is intended to replace the maternal relationship, the fear of the unavailability of the peer makes a healthy relationship impossible.

Children generally need two parents for the development of a healthy ego. The loss of a father because of death or divorce is highly traumatic. As we have seen, however, total absence is not necessary for a father to be unavailable to the child. Many more fathers are engrossed in their work and in their own needs, to the exclusion of their children, than are physically absent from the home. They feel they are meeting their responsibility because they are devoting time and effort to their work to support their family. Nonetheless, this is often only a rationalization for diverting their involvement from the home, as many are in a financial position to be able to spend more time with the children.

The type of absence we have called being "distant" frequently produces a particular form of pathology in the child. The child is confronted with a parent who is there, but who isn't there. A highly ambivalent feeling toward the father results. Even when the father is home, he is often absent from the family. Dinners are often eaten separately, and the mutual watching of television provides the only mode of communication between the father and the children. The resentment that the child harbors against this type of man is intense, yet he constantly hears the remark, "Daddy is tired and has worked hard all day," and therefore the resentment becomes unconscionable. The children turn to the mother for both mothering and fathering. The result is confused identity for the son and suppressed resentment towards men by the daughter.

Even though the lack of involvement with a child is sometimes very subtle, the child is aware of it. On an unconscious level he senses that the parent is not involved. Anxiety and resentment develop. The anxiety is paramount as the feelings of security in the parent-child relationship are absent. The resentment heightens the anxiety because the child is fearful of losing what minimal involvement he does feel with the family.

It has been the experience at the day care centers with which I am involved that this family structure is the most common forerunner of the drug-abuse personality. It is the subtle emotional loss of the parent, rather than the actual physical loss, that creates the greater problem. It is this narcissistic emotional absenteeism that is the breeding ground of the pseudo-independent dependent personality that so frequently takes drugs for relief.

—4—
The Inadequate Parent

There are numerous other neurotic family patterns that can be precursors of drug abuse. One of the most common is that of the inadequate parent. Unlike the narcissistic parent, this parent is actually invested in the child's needs but, because of his own neurosis, he is actually emotionally incapable of performing his duties as a parent. Frequently he feels genuine remorse at not being able to fulfill the role he knows is necessary for his child. Intellectually he generally understands his role as a parent, and it is only emotionally that he is incapable of acting upon what he knows.

Diagnosis of such parental personalities reveals obsessive/compulsive and phobic neurosis, inadequate personality, anxiety neurosis and various other severe personality problems that an individual develops. It would be repetitious and unnecessary to single out each of the neuroses and how it affects the children, for there is a remarkable similarity between the feelings of absence that the child experiences with a neurotic parent, and the feelings of absence that he experiences with a narcissistic parent. One of the major differences between the two types is that the neurotic parent is aware of his lack and is experiencing a sense of discomfort which will frequently cause him to seek help for himself or for the children. In the narcissist, the converse is true; he may have totally justified his mode of existence to himself and he experiences no discomfort at the thought of what he is doing. He is loathe to seek help for himself, and is equally loathe to seek help for the children, who are seen as an extension of the self.

It is exceedingly difficult for a mother who is suffering from her own neurotic symptoms to give the love and affection and caring that the child demands. She is too emotionally tired to meet the child's demands, and often too fearful of her response to deal with the child in a comfortable manner.

A tense mother can produce a tense infant. Studies have shown that the maternal holding of the child from the day of birth can influence the child's degree of tension. The mother who from the beginning is anxious and fearful about holding her baby can experience severe muscle tensions causing a similar muscular response in the child, while the mother who is relaxed from the beginning will transmit the relaxation to the child. This fear of dealing with the child remains throughout the mother's life. As each minor crisis arises for which the child turns to his mother, the mother can only respond with anxiety often bordering on panic. This anxiety is sensed by the child and what is essentially a minor trauma becomes a major emotional crisis. The child fails to learn that these minor traumas are an everyday occurrence in a person's life, so he becomes fearful that they will cause overwhelming destruction and unhappiness. Frequently the child then attempts to avoid any traumatic experiences at all. Then comfort-producing drugs become an easy way out.

Extreme anxiety in handling the day-to-day occurrences is frequently seen in the phobic, the anxiety-ridden, and the inadequate personality. The obsessive/compulsive individual, though, is one who has a great deal of difficulty in demonstrating any type of emotion and tends to deal with all crises, internal and external, intellectually. A large percentage of our society is based on some obsessive/compulsive character traits; indeed, it would be difficult for any of us to function in this society without them. Nonetheless, when these traits become excessive they become a hindrance rather than an asset. The child experiences his parents as being cold, remote and unfeeling and often sees himself to blame. He does not see it as a parental shortcoming, but as his own shortcoming, that prevents him from receiving love and security from his parents.

Emotionalism is a very important part of rearing children. It is an experience of the body as well as of the mind. So tactile sensations are quite necessary. A warm hug and a kiss to a child are of utmost importance, as they produce the feelings of love. It is very difficult for the obsessive/compulsive neurotic individual to show this kind of affection because of his own anxieties and fears; affection is therefore lacking in the child's early life. The result is a nonemotional individual. The child in turn may identify with the parent and develop his own obsessive/compulsive character defenses, or the child may not be strong enough to use this type of defense and will revert to a more permeable defensive structure, leaving him fraught with anxieties. As has been mentioned in the first section of this chapter, a child who is going through even normal anxiety may turn to drugs in order to alleviate the anxiety and produce comfort. The lack of affection from an obsessive/compulsive parent creates an anxiety in the child which is greater than normal, and therefore the resulting use of drugs will not tend to be self-limiting but rather self-perpetuating.

— 5 —
The Psychotic Parent

Far more devastating to a child than a neurotic relationship is a psychotic family relationship. Here one or the other parent is not in touch with reality and deals with the world through fantasy. Parental consistency or affection is absent and the child experiences intense discomfort from his parents. Because of the dread of any relationship that is present in a psychotic, it is difficult for him to display any affection or to become close to another person, including his children. Much of the dread and the fantasy life is transmitted to the child, and often a child, who has realistic awareness of the world about him, is unsure of his own perceptions because they differ from those of the psychotic parent. The child has to be reassured in later life that it is he and not his parents who is seeing the world realistically. As it is from the parents that a child first develops a sense of how to cope with the environment, he becomes greatly confused.

Psychotic interrelationships take many forms. Incest is frequently seen, with all of the trauma that incestuous relationships can develop. Sibling incest seems to be quite common in the homes of psychotic parents. Sibling incest is such a highly emotionally charged situation that neither child can cope with his or her intense erotic feelings, which often remain for the rest of his or her life. Yet while they are highly eroticized, there is also an overwhelming sense of guilt that is brought about by this socially taboo relationship.

Frequently there is uncontrolled violence in a psychotic family. The father and mother may physically attack each other with such diverse weapons as knives, pots and pans, or anything else with which they might cause harm to one another. Overt violence causes tremendous trauma in the child, who is frequently a witness to these upheavals. The loss of control that is experienced by the parents is perceived by the child as a normal condition in highly emotional situations; the child therefore becomes unsure of his own sense of control.

The psychotic individual is far more threatened by his own internal world than he is by the world at large. The children of psychotic individuals develop the same fears. Witnessing uncontrolled rage makes the child extremely fearful of his own anger. Yet anger grows more intense in these children because of the repeated frustrations they experience. They are usually afraid to express any anger or resentment towards the parent or towards the world at large. They see the situation as one in which they too can lose control and destroy whomever they are angry at. It is nearly impossible for children of psychotic parents to grow up emotionally stable, and even prior to the age associated with drug abuse, these children often manifest severe emotional problems.

As the whole drug scene is such an efficient defense mechanism against both internally and externally produced anxiety, and because of the easy availability of many kinds of drugs, they become the most common method of warding off the anxieties that could have been handled in many other ways—as they were prior to the popularity of drug use. It is not then a direct cause and effect relationship between a neurotic or psychotic family and drug abuse. Rather, the family structure leads to emotional disturbance. It is the emotional disturbance which causes the individual to seek comfort. Then it is the chemical comfort of drugs, the subcultural acceptability of drug use, and the interpersonal relationships that develop within the drug culture, which, taken together, make this such a desirable mode of escape from anxiety.

—6—
Our Drug Oriented Society

Drugs do not thrust themselves upon an individual; an individual must seek them. How then does our society play such a tremendous role in the present epidemic of drug abuse? Why is the American society more prone to the use of drugs than is either the Oriental or European society of the 1960's and 1970's? Very few societies in the world are as comfort oriented as our own. In all societies one can find many individuals whose primary goal is the comfort of their own bodies, but it is more frequent in America. Comfort, however, must not be confused with enjoyment.

Many enjoyable activities are pursued despite discomfort. Among the most common are skiing, sailing, scuba diving, mountain climbing, and sky diving, where there is obvious sacrifice of comfort for enjoyment. Recently, a group of young men from the New York Explorer's Club set out to climb the sixth highest mountain in the world. They were willing to face 50 knot winds at minus 50 degrees Farenheit, hanging at the side of a cliff on a rope attached to a little piton. To them, this was enjoyment but not comfort. These men obviously constitute a minority in our society. The American dollar is more readily spent on air conditioning, color television, luxury hotels, and fast plane travel than it is on hiking boots and skis. The majority of the cars on the road are equipped with power steering, power brakes, power windows, power seats—most of which increase the danger of driving as they allow the individual to become totally relaxed and almost to forget what he is doing.

The primary emotion that is caused by drug abuse is not thrills, but comfort. It makes the individual forget his cares and worries. It makes the reality of the world unimportant and unnecessary to cope with.

It is unfortunate, but much of the blame can be placed upon

the scientific discoveries of the past forty years in the areas of medicine, and the use and misuse of drugs by doctors and patients throughout this period. In the 1930's penicillin was discovered. This was followed by a rash of "wonder drugs" that cured most bacterial infections. No one can deny the greatness of the discovery of these wonder drugs. Nonetheless, misuse rapidly followed. People were no longer satisfied with receiving a simple aspirin to help overcome the symptoms of a common cold. They expected a drug and a miracle to help with every illness.

As an emergency room physician, I frequently encountered adults who demanded antibiotics for a cold, despite the fact that it was of viral origin. They had heard, of course, how these drugs can cure the common cold, but were ignorant of the fact that they have no effect on a virus. Parents often demanded medicine for their children not so much so that the child could sleep but so the *parent* could sleep. I have never been opposed to allowing a child more comfort and sleep during an illness, but the demands of a parent for a drug that I consider deleterious in order to allow the parent to sleep are, to me, totally irrational. These incidents exemplify early steps toward drug abuse. The parents' demand for antibiotics for the child has often been called the "plight of the pediatrician." More than one pediatrician has had a patient turn from him to another doctor because he refused to supply a useless drug to the child.

More closely related to the problem that we now see in drug abuse was the discovery of tranquilizers. Tranquilizers have, of course, markedly altered our present state mental hospital system for the better. Many individuals who would have spent the rest of their lives sitting in a mental hospital are functioning and performing useful tasks with a sense of self-esteem and self-respect. But, once again, the abuse has become rampant. Any individual can turn on a television set and be directed to the nearest drug store to supply him with any one of a number of agents to help him cope with his daily tensions and anxieties. Advertisements depict a young mother transformed from a screaming, threatening, punitive authoritarian figure to a loving, warm, concerned parent—just by taking a simple pill to calm her nerves. We see a harried businessman, on the verge of losing his job because of emotional strain, suddenly being promoted to a better position because he had a good night's sleep, thanks to one of the numerous sleeping potions sold over the counter.

Aside from legal and moral considerations, it is impossible to differentiate this kind of self-medication from the drug abuse that is so roundly condemned by the very same parents. Diet and sleeping pills prescribed by doctors have become the mainstay of many medicine cabinets. It is no longer deemed necessary to have will power to lose weight; one merely takes a pill, which is usually composed of an amphetamine and a barbiturate in combination. If will power is no longer necessary to lose weight, why should it be necessary to stick out a particularly dull class at school. It is much easier to take that very same diet pill and enter the classroom in a state of oblivion than attempt to listen to a boring teacher drone on in monotone and to make some sense out of what he is trying to teach.

The influence of parents using this kind of medication is twofold upon the youngsters. Primarily it gives them a model for identification. Children tend to imitate parents both in characteristics deemed desirable and those deemed undesirable. Secondarily, many a youngster gets his first supply of drugs right in his own medicine cabinet. He knows the contents of the "Dexamils," the "Obitrols," the seconals, and the nembutals that are being used by the parent.

Let us also recognize the hypocrisy of the parent who uses these drugs, and then admonishes his child to refrain from drug abuse. The child sees the hypocrisy and disregards the parent's advice. Far more overt is the use of alcohol and nicotine. These drugs, used in excess, are as dangerous as many of the drugs that adolescents abuse. Parents serve as a role model: when a child sees a parent in a state of alcoholic intoxication, he forms a vision of himself. A parent who is a chain smoker and expounds on the dangers of marijuana carries very little authority with any intelligent adolescent.

Basically, then, there is an attitude that permeates a society which makes the use of drugs acceptable. There is a clear dual standard which is observed by most adolescents and is considered hypocritical. They know that there is no difference between a diet pill and an "up," a sleeping pill and a "down." They know that marijuana is less physically dangerous than nicotine.

They know that people smoke to obtain comfort and relieve tension. Why not use marijuana in the same manner? They see their parents have a drink before dinner; they know that marijuana and alcohol have similar effects. What rationale can be given to a child not to turn to harder drugs?

The "work ethic," which was so basic to our forefathers, has been completely submerged by the immense effort an individual now puts into finding more leisure time to waste. Leisure time and comfort seem to be the two goals of the American dollar. The struggling which is frequently necessary to achieve any worthwhile goal is looked upon with scorn and disdain. The man with the 100-foot yacht is looked upon with more admiration than the scientist or the educator who, because of a poor business sense, or dedication to his science or his education, ends up without the profit of his discoveries. On the contrary, a sense of pride and a sense of achievement, both of which make the turning to drugs less attractive, comes only through struggling. The more it is necessary to struggle towards a goal, the greater meaning that goal has for the individual. Those men struggling up a Himalayan peak would get very little satisfaction, gratification, or pride from walking up a hill, though to be sure the struggle would be far less and the comfort far greater.

On the other end of this spectrum is the individual who exists for 90 percent of his life in the standard society of western culture. However, he is constantly under the influence of some form of self-medication. The medicine that he uses may be amphetamines, in the form of diet pills, to wake him up in the morning and allow him to function during the day, or sleeping pills, which allow him an uninterrupted rest at night. Frequently, the amphetamines and soporifics are mixed up: he will awake in the morning, take a "downer" and go off to work in a daze; come home in the evening, take an "upper," and fail to fall asleep.

There is also the alcoholic who is able to maintain his job, but who will regularly have a cocktail or two at lunch, at dinner, and before he retires. Without his drink he finds that he cannot face the daily stress of the business world. Though these people constitute the vast majority of drug users in this country, they are less obvious to the casual observer.

In the late 1950's and the early 1960's the world became

aware of the "hippie" coming from a bohemian subculture that consisted mainly of individuals who generally were creative in artistic fields. The "hippie" aspired to the life of writers, painters, and playwrights. The use of marijuana was prevalent in the bohemian-hippie life style, but alcohol and hard drugs were unusual. Out of this group, a number of young writers developed, the most famous of whom was Jack Kerouac. They advocated "dropping out" of the materialistic society and entering a life style totally opposite to that of the productive individual. Kerouac became the standard bearer for a large group of young people who felt that they did not have to cope with the stresses of American society.

At this time there emerged a more respectable spokesman for the counterculture, Harvard professor Timothy Leary, who advocated the use of a more complex drug, LSD, as a mind expander. The two philosophies nevertheless went hand in hand. Leary had taken many "trips" on "acid," and became the idol of many youngsters in crusading for his discovery. As this subculture grew in numbers, it was easy for young people to drop out of their homes and find a place of refuge. They were generally dissatisfied, unfulfilled, and unable to exist comfortably in society anyway. They generally had difficulty in relating to their parents and to their peers, and found in the hippie subculture a family substitute.

In their counterculture world they felt accepted and wanted because few demands were placed upon them. Many survived by begging in the streets; others took meaningless jobs as sales people in the numerous "head shops" that sprang up, or as waiters or waitresses in the small restaurants in the "hip" community. When they tired of this they left without warning, and merely drifted away, either back to their "pads" or across the country. It seemed that there was an entire generation of young people on the move; as one drove along the roads he would see many bearded, sandaled, long-haired young people, hitching rides from here to there, but not really going anywhere.

The only demands placed on these people by their group were to renounce materialistic values, to be involved with drugs, and to be willing to abandon the values with which they had grown up, including education and sexual inhibitions. Many youngsters claimed that this way of life allowed them to become individuals. They claimed that their elders had handed them a society that they could not tolerate, a society full of wars, poverty, hunger, and man's inhumanity to man. While they were close to the truth, they had no solution but to drop out and to make no attempt to alter the world situation they had inherited.

The group offered a new way of life and a new value system, which supported the individual's rationalizations and denials. To function in this new life style, one would pretend that things were the way he wanted them to be. One of his major denials was that of the danger of drugs. Members of the group would reinforce this among themselves, claiming that they knew just how much one could take without running the risk of harm. When they would witness a death by overdose, they would claim that he either received "bad stuff" or that he did not know what he was doing. The group assured them that what they were doing was right, and that some day people would see that their mode of existence was preferable. The lack of productivity, the lack of self-esteem, and guilt about the sexual acting out that so often accompanied this life style required the individual to continue on drugs to relieve his anxiety.

New standards, values, and heroes emerged in the hippie subculture. Timothy Leary, Allen Ginsberg and Jack Kerouac became almost legendary demigods to an ever-growing group of disciples. New musical groups, such as The Beatles and The Rolling Stones, idealized this life style and the use of drugs on an epidemic scale. Many song titles were thinly disguised terms for drugs. *The Yellow Submarine*, the film done so beautifully by The Beatles, was a poetic depiction of a drug trip, and the title itself a reference to a drug capsule. Musicians most popular with the young set openly displayed their own drug abuse.

It is true, of course, that jazz musicians especially have used drugs since the turn of the century; but popular performers have never been so proud of it and open about it as in the last fifteen years. As "rock" heroes died from overdoses, they were venerated by youngsters.

In the subculture a new hierarchy was developing. Heroin addicts, acid heads, down freaks, and speed freaks were hip; marijuana and alcohol were looked upon as something for the "teeny boppers." To what length one was able to grow one's hair became a symbol of superiority. The number of trips a person had on LSD, the more "way out" he appeared to his friends, the more bizarre the dress, the more downs that he was able to swallow, and the more speed that he was able to shoot became symbolic of leadership and "greatness."

In the addict society, made up of hardened junkies, there was an even more clear-cut negative value system. A person who had been convicted of a petty misdemeanor boasted to his friends that he had been arrested on a felony charge; a simple shoplifting conviction was often claimed to be grand larceny, and an arrest for possession was turned into an arrest for dealing a kilo of heroin. To promote his status, the junkie exaggerated the amount of time he spent in jail to his friends.

It was into this society that the drug abuser fled. He found refuge and support along with drugs in the group. As all men need group support, it was hard for people trying to break through to this youngster to make a convincing case. The group allowed him to avoid the older generation for support, and in most cases he actively turned away from anyone whom he saw in a "straight" society. The watchword had become, "Don't trust anyone over thirty," a sentiment that has become a joke in only a few years.

Most prevalent, but less obvious, was the fringe society of drug abusers. These were the youngsters who remained at home, continued in school, and maintained some of their old peer relationships, but at the same time were heavily involved in the drug culture. Their parents and often their straight friends were unaware of their drug use. It was easy to pick them out as being outside the mainstream of society, but hard to distinguish them from those nondrug users whose political and ideological philosophies also placed them apart. Long hair and old army jackets were their uniform. With drug users, however, one also saw a marked degree of truancy and class-cutting in school, a falling of grades, a withdrawal from parents and old friends, and formation of new relationships in drug oriented groups. The most commonly abused drugs were marijuana, hashish and barbiturates, with LSD decining in the 70's as the frightening consequences of this drug were publicized. Heroin and methadone are now used with ever-increasing frequency in white, middle class suburban communities.

Youngsters in this fringe society usually get high during the day when they are supposed to be at school, remain in class in a state of inebriation on drugs, and then go home, relatively straight, often to get high again before going to bed. Their parents are rather oblivious to this pattern and usually only notice a fall in marks and a report from the attendance officer at school. If parents should come across the youngster's cache of drugs, they are all too ready to believe their child's explanation, "I was only keeping them here for a friend and I'll get rid of them." Parents, like youngsters, are ready to deny the reality of the situation in order to avoid unpleasant confrontations and internal turmoil. Parents might find their youngster's antisocial behavior unacceptable, and the resultant fights might culminate in the youngster's running away from home for a few days.

This fringe group usually would not advocate dropping out of society, though many of their values are held in concert with their more rebellious colleagues. A few adolescents who remain on the fringe profess a desire to attend college and eventually enter the mainstream of life.

The fringe culture has its own standards of values. There is a strong bond of loyalty between the dealer and his customers. A closed group, it might be a cross-section of school athletes, who not infrequently play their games under the influence of marijuana, of better students, whose drug use is limited to evenings and weekends, and of the bottom of the academic and social ladder in high school society.

Relationships between two individuals are less common than a relationship to the group as a whole. Best friends are often changed on a weekly basis. A boyfriend-girlfriend relationship is often looked upon with scorn, and sexual behavior is often carried out among all the members of the group. Promiscuity in girls in the group is common. To be unwilling to perform sexually is considered a "hangup" of the straight world. Nevertheless, many girls maintain their parental values and morals about sex on an unconscious level. This leads to strong guilt feelings, which are once again repressed and denied. It becomes necessary to maintain themselves on barbiturates in order to keep their guilt on an unconscious level.

We can see now that this fringe society has a few simple motivations. Drugs are the key to a wonderland that they feel they could not otherwise witness. Sustained use produces a tranquility and peace that is unknown to youngsters. Their central value has been changed from productivity to drugs, so they no longer have to feel anxious about school, friends, and parental expectations. Most have never felt comfortable in a school setting; expectations from the school and from their parents are more than they can tolerate. Most have poor peer relationships even prior to entering the drug culture. Their relationships are minimal and never very intense. They are threatened by a relationship with members of the opposite sex and constantly anticipate rejection and ridicule. Even the ones who are involved in athletics and academic work feel separate and isolated from their peers.

In more than three hundred interviews with youngsters in this group, I have rarely found an individual who felt confident in his ability to achieve his goals and confident in his relationships with his peers. They would often state that on the surface they felt and tried to show they had friends, but internally they always anticipated that their friends did not like them and were being false. They believed it was dangerous to form a one-to-one relationship with someone of the opposite sex because of the risk of rejection. Even those youngsters who functioned well in a school situation always had the apprehension that they could not continue to perform the way they had in the past.

A sense of insecurity pervades the drug abuser's thoughts. It is only in very rare instances that an individual who is abusing drugs does not have a severe emotional disturbance, that would have been discovered if he had been thoroughly evaluated prior to his drug involvement. Only an unusually traumatic circumstance, such as a war experience, precipitates drug use in a youngster who is not otherwise disturbed.

The final group of drug abusers consists of those who are very much in the mainstream of society. Businessmen, workers, and housewives become involved for other reasons than the youngster does. These people have not "dropped out" from society, and rebellion against the values of our culture is the farthest thing from their minds. Rather, they find themselves insidiously hooked on amphetamines and barbiturates, tranquilizers, over-the-counter sleeping pills, and tension-relievers. Even ordinary aspirin has become an abused drug to some degree.

Such users are differentiated from the others for the simple reason that they are ashamed that they are using drugs. They do not tend to reinforce one another, as they keep their drug use secret from even their closest friends and relatives. Yet they do have a profound effect on the adolescent drug abuser, in the sense of identificaton of the child with the parent.

Most children learn, at one time or another, that a parent is abusing drugs. They will find the parent demonstrating the symptoms of drug use, or they may actually find drugs in their parents' medicine chest. Many youngsters have indeed gotten their start by taking drugs right out of their parents' closets. These same parents are often hesitant to end their drug involvement when their children are in a therapeutic situation. In this way they become similar to their children, in that their involvement in drugs supersedes all their other standards. They are able to rationalize their behavior by drawing distinctions between various types of drugs. As we shall see, many of these distinctions are fallacious.

A second group of adults involved in the drug scene are alcoholics. A large percentage of drug-abusing children come from families where one or both parents are alcoholics? Hard as this fact is to accept by parents, the child sees no difference between the abuse of alcohol and the abuse of drugs. These adults do little to hide their drinking habits. The child often sees them display their antics at cocktail parties or even in the home after a "hard day's work."

One young man whom I recently interviewed recalls with horror a mother who was drunk every day and the frequent violent bouts that she would have, flinging whiskey bottles and demanding more. It is hardly surprising that this individual would turn to drugs. Parents tend to forget the very basic fact that they act as a role model for their children; they cannot blame society for the shortcomings of their children. If a parent does not want a child to imitate his behavior, he had better make sure his child has no knowledge that the parent is acting out in this manner. I doubt this is possible. Yet many cocktail parties are given blatantly in the presence of youngsters in the house.

Recently, more and more of the older generation have been smoking marijuana. In order to prove to his youngster that he is trying to understand the new generation, a parent frequently tries "turning on" with his children. What does a youngster think of a parent who does this? Usually the child is amazed and astounded. He feels infringed upon, that his parent is behaving in a way that he cannot be proud of. Despite frequent, overt protestations to the contrary, in which a child expresses openly that he feels his parent is "cool," after deeper exploration the child will admit loss of respect and embarrassment at his parent's behavior.

Most of the older generation, fortunately, have the discretion to keep their marijuana smoking well hidden, but it is not unusual, though, to find smoking going on right at the same cocktail parties where the children are awake upstairs. Such behavior may be part of a trend for the older generation to emulate the younger instead of vice versa. Though we would like to pass the blame onto adolescent peer groups for the drug society, it is we who have taught this generation the uses and excesses of a chemical reaction to our psychological problems.

—Part Two—

Storm Warnings

What children cry out for is seldom what we listen to.

—7—
Examples Parents Set

We have now come to the point where we must accept the fact that there is a causal relationship between a drug-abusing parent and a drug-abusing youngster. Far more adults abuse drugs than one imagines. Parental drug abuse can take many forms: nicotine abuse in the heavy cigarette smoker, alcohol abuse in the heavy drinker, abuse of prescription medicines or marijuana abuse which differs little from that of the adolescent. The common parental drug abuse of alcohol and two-pack-a-day cigarette smoking may not be a direct cause of a child's drug abuse, but the child becomes aware that the parents have rationalized away a dangerous habit. He is, therefore, not at all swayed by the argument that his use of drugs is dangerous to his health; the rationale of the danger of drug use is unconsciously countered by the child. Nicotine abuse differs from alcoholism at least in that the latter is sufficient to cause a child to become a drug abuser. There are very few alcoholics who have no other emotional problems aside from their indulgence of alcohol, but what the child *witnesses* is the use of alcohol.

Heavy drinkers give the child an unhealthy model with which to identify. The model is of a parent who is incapable of coping with internal tensions and anxieties and the anxieties produced by the environment. He or she has turned to a drug to alleviate these tensions. Most children currently are convinced that alcohol is a more dangerous drug than marijuana, so they often turn to marijuana with the rationale that they are not doing anything nearly so bad as their parents. The parent rationalizes his or her use of alcohol to the same degree as the child rationalizes his or her use of marijuana. The incontrovertible argument to the child is that alcohol is legal, but the child will inevitably counter, "Well, marijuana should be legalized."

The alcoholic parent is emotionally removed from the child and emotionally attached to the bottle. It is a highly infantile form of existence, one that precludes a mature relationship which a child expects from his parents. Family ties are weakened in an alcoholic home and family stability is generally lacking.

An alcoholic is not necessarily a drunkard; in fact, many alcoholics rarely appear inebriated. An alcoholic may take two or more drinks a day at lunch, before dinner, and prior to retiring. With this form of alcoholism there is a subtle relationship between the parent and his children. It is primarily by identification that the child turns to drug use when he has an alcoholic parent, but coincidentally there is also a need for revenge towards the parent. The child functions on the primitive level for revenge of "an eye for an eye" and unconsciously is stating in his actions that if the parent is going to abuse alcohol then he is going to get back at the parent by abusing drugs. In a later chapter, the role of drugs as a weapon against the parent will be discussed more fully, but let it suffice to say here that the weapon against the alcoholic parent is appropriate in a child's mind.

It seems to be becoming increasingly fashionable for parents to "turn on" with their children. Modern, liberated parents are frequently smoking marijuana and more and more in the company of their children. This is a highly infantile reaction on the part of the parents. It goes beyond condoning a dangerous action that is beyond the parents' control to actually supporting such action. It supports the child's illegal behavior and self-destructive attitude. Often it is resented by the child; a number of children have stated that they resent their parents' acting like adolescents, and would respect them far more if they condemned the practice of marijuana smoking rather than indulging in the practice themselves. Adolescents know that smoking marijuana is an illegal and rebellious act and they see it as acceptable in the adolescent years. They do not, however, condone it as adult behavior and many of them are fully aware that they will give up the practice once they reach adult life.

Many adults are habituated to marijuana. The effect that it has on the children is similar to the effect that parental abuse of alcohol has on them, except that the use by the parent is far more clandestine than is alcohol use. Because the illegality necessitates hiding the use, the child may not be as fully aware of what the parent is doing. However, children will often find the implements of smoking, such as the rolling papers, marijuana seed or "roach holders," in the parents' possession. It is very hard for parents to condemn the practice in a child when they are indulging in the same activity. "Do as I say, not as I do," is a feeble excuse. The habituated marijuana user, like the alcoholic, is an emotionally disturbed individual, and the effects of the emotional disturbance will cause symptoms in the child. The use of marijuana by the parent substantially increases the chance of the child using the same escape mechanism for his problems.

Many "legal" drugs have appeared on the market in recent years that are being abused by adults. They include over-the-counter tranquilizers, sleeping pills, and diet pills that the adult takes with little thought of the effect it has on the child. They are justified, of course, in terms of adult tensions, the need for relaxation, and the inability to sleep. Adults seem to have forgotten that adolescents may have more anxiety than they do.

The crutch which drugs represent is supposed to help an individual to cope, but it also has the side effect of increasing the chance of the child's becoming a drug abuser. Many of the tranquilizers prescribed by doctors are a necessity. Yet most children do not make the differentiation in their own minds between the uses of drugs. They see all drugs, whether over the counter or by prescription, as being similar to their own use of barbiturates and amphetamines.

Barbiturate and amphetamine use is also prevalent among adults. Barbiturates are, perhaps, the drug abused most commonly by adults, aside from alcohol and nicotine. Adults tend to use them in the evening as an aid to sleep, but the quantity is often far in excess of a safe prescription. The body's tolerance grows, and it is often necessary to take three or four pills in order to obtain the desired effect.

Children commonly confess that their first source of supply of drugs was their parents' medicine cabinet. Once again, the parent is under the misconception that the child is unaware of what the parent is doing. *It is only in extremely rare instances that a child is unaware of the parents' activities, no matter how well the parents try to conceal them.* A child is a curious creature and at certain ages will tend to explore whatever is behind closed doors. He soon becomes aware of the parents' supply of medicines. In addition, the process of identification comes into play and the child emulates every action of the parent.

The emotional disturbance in the parent develops the matrix of the emotional disturbance in the youngster. It is the identification with the parent that causes the youngster to cope with his emotional distress as does the parent through the use of drugs. Accordingly, *any* use of drugs by the parent can cause the child to turn to drugs.

In all of the cases we have seen so far, there is one thing in common—the parent with an emotional problem. It is also possible for a *family* group, however, to have a neurotic interaction, with no emotional problem evident in the parents. The interaction itself becomes the causative factor in the child's seeking an escape in the drug society. It is not that the parents or the child have no neurotic conflicts individually, but that their internal conflicts are not as severe as would warrant the severity of the reaction. What happens is that the whole becomes more than the sum of the parts.

This type of situation is becoming increasingly common as the two opposing cultures of the younger and older age groups become increasingly diverse. It has been a manifestation of our society throughout recent years for the younger generation to branch off more and more from the family tree and seek new roots that are alien to their parents. The change is often influenced by the advanced education of the child and the increased contact that the youngster may have with different cultures. It is not unusual for children under the age of twenty to be touring foreign countries and different areas in our own country, broadening horizons and experiences beyond what was available to their parents. It is difficult to call it a symptomatic disorder when parents in their fifties who are set in their ways find it hard to change as the child comes into direct conflict with his parents' ways. The loosening of family ties makes identification with his parents difficult.

It may be argued that, if the early relationship between parent and child has been solid, the love relationship would be strong enough to overcome any situation. This is probably true. In most occurrences a breakdown comes in a borderline relationship that would ordinarily be strong enough to withstand the normal trials of an adolescent's growth. Now it is suddenly taxed beyond its means.

Another situational factor that might divert a child from family ties is the experience of war. A young man is sent off to the army and comes back in a state that is entirely different from when he left. It was not infrequent, following the recent Vietnam conflict, to find many veterans markedly changed either towards the side of pacificism or towards the side of increased aggression from the time that they arrived in the war zone. This experience too can tax a relationship beyond that which is ordinarily encountered and may tax an individual to such a degree that he will turn to drugs.

Many veterans returning from the war felt that it would be impossible for anyone who had not experienced the turmoils that they coped with to really understand the stresses of their lives. It was understandably difficult for them to return contented to the society that had been at peace while they had been risking their lives at the front line. The family structure no longer served a vital function in their lives. As close as they might have once been to their parents, they found that the parental relationship no longer fulfilled what they were seeking. Their buddies in the war were dispersed all over the United States; suddenly they found themselves in an isolated position, feeling out of step and unequipped to deal with a peacetime society.

Many veterans started to use drugs while abroad in order to help cope with the constant tensions and fears that one experiences in the war zone. In this conflict especially, they were confused as to their duties and their mores. They were not involved in a war that was clear in its design, its morality, and its purpose. They did not have the solid backing of the people at home, as in previous wars, and they themselves often were in serious conflict as to the morality of the job that they were doing. They could not arouse in themselves the anger that is necessary to cope with the stress of the risk of being killed. They therefore lived under a far greater fear of death than was witnessed in previous conflicts.

It is now common knowledge that many of the soldiers returning from Vietnam were addicted to heroin, and many more were regularly using marijuana and hashish. Fortunately, a high percentage of those addicted have been able to end their habit on return to the United States. War causes a traumatic neurosis which is usually self-limiting as the stress is removed. It can be interpreted that the high percentage of those who were able to withdraw from drugs was due to the diminishing need as the stress of war abated. These men evidently did not have preaddictive personalities; the drug-abuse situation was temporary, ending when the anxiety from the overwhelming trauma ended.

Other situations that may cause trauma to a child can arise when parents have been misguided by people to whom they have turned for advice regarding their child's education. It is not unusual to find well-meaning parents who would be willing to accept lesser goals for their child, but have been urged by guidance counselors or school officials to encourage their child to further their education. Often these children are quite capable intellectually of coping with the experience of college but, because of emotional immaturity, cannot cope with the challenge. Nonetheless, the parents urge the child to try to deal with situations that cause frustration, feelings of inadequacy, and subsequent anger at the whole situation. This child is not the product of an unloving home, but the product of an unaware educational system.

In many similar situations, the individual case can be understood if it is recalled that *drug abuse is a defensive mechanism against anxiety*. It is easy to understand how anxiety is provoked by the lack of support that the individual feels from his close associations. Considering what has been said previously about the mechanisms of drug abuse, it follows that the tranquilizing effect of drugs becomes a welcome escape from the feeling of inability to cope with those situations in which individuals find themselves. As the whole drug culture serves to hide symptoms behind the behavior cover, it is often difficult to make a clear diagnosis. In a recent address to the Nassau Psychiatric Society, Jacob Arlo noted similar factors and pointed out that we have to look behind the commonly accepted symptoms to make more accurate diagnoses. It is often necessary to limit his drug use so that the adolescent can become more aware of his underlying feelings before any diagnosis or awareness of the problem becomes apparent.

—8—
Signals from the Child

REBELLION

There is a vast difference between what one sees on the surface and the actual character of the drug abuser, addicted or not. Most of the popular pictures and legends of drug abuse are based on superficial considerations. The beliefs of the general public are also the basis for many of the drug laws that have arisen in the last 75 years. It is important to understand the true character of the drug abuser in order to determine what treatment is necessary for him and what laws can be enacted to prevent his misery from becoming epidemic.

The Nonaddicted Drug Abuser

On the surface, the nonaddicted drug abuser presents the picture of a merely behavioral disorder. "Behavioral disorder" is a term used to describe a multitude of symptoms that make their appearance in an individual's life. The symptoms are exhibited in drug abusers, alcoholics, truants, delinquents, chronic criminals and anyone who manifests an antisocial behavior pattern. "Antisocial" behavior is defined as that which does not take account of the needs of society at large. It is characterized by a lack of conscience or consideration for people. In the nonaddicted user, perhaps the outstanding symptom is rebellion, which is primarily directed against the individual's family.

It manifests itself in frequent fights and arguments—which usually bring the family to seek help, unless drugs have not been used as the alternative. These fights, not infrequently, become violent; it is not unusual to find individuals in a family getting into violent demonstrations nightly. Most of the time

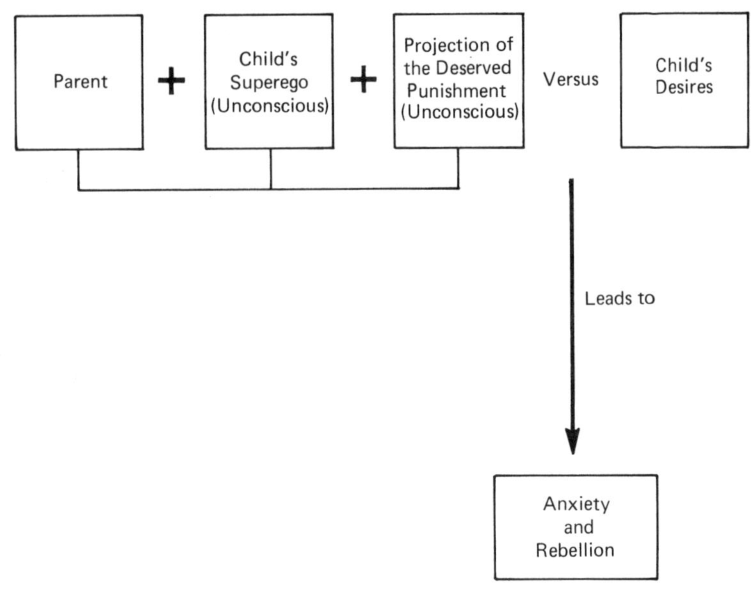

External Conflicts

arguments spring from a child's demands. Very frequently these demands are totally outlandish: to remain out all night when he has school the next day; to be allowed to take off from school to pursue a sudden objective; to quit school entirely. Illegal truancy from school quickly comes to the attention of the parents along with other violations which conflict with the parents' standards.

Often, however, familial fights develop from the outlandish demands of the parents. It is a common occurrence for parents to forget the needs of their child, or to make demands that are entirely different from what can be expected of his peer group. He may be asked to come in at eight o'clock when the rest of his friends have a ten o'clock curfew. He may be forbidden to make "phone calls," or speak to people of the opposite sex. He may be asked to miss important events, such as parties, dates, and concerts to which his friends are allowed to go. Fighting follows which is different from the usual and normal arguments that evolve in any household in which an adolescent lives: it is more frequent and more intense. There is a constant struggle between parent and child to gain mastery of the situa-

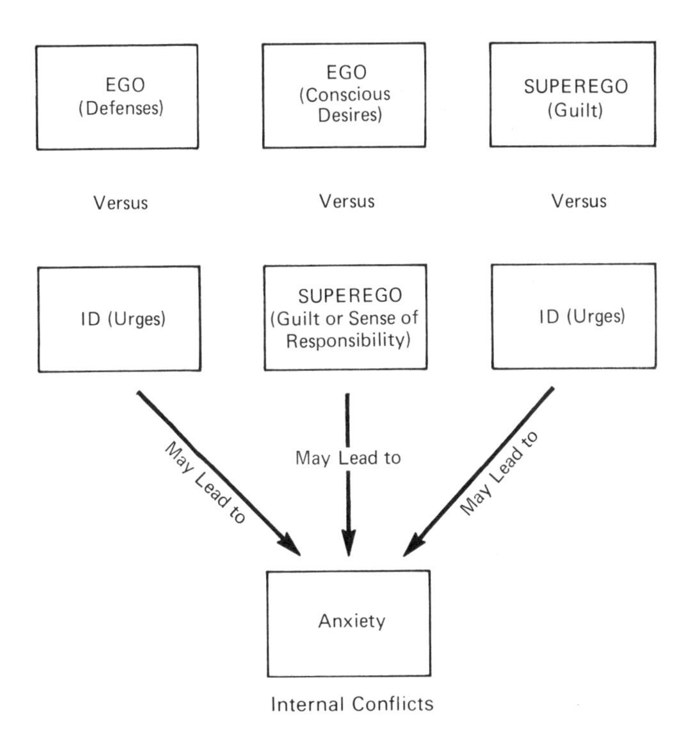

Internal Conflicts

tion. The struggle leads to total isolation of the child, who becomes unwilling to participate in any family matters or pursuits. He withdraws to his room, locks the door and turns on his music. He may walk out of the house without permission and return late at night, much to the distress of his parents. If the fighting gets particularly intense he may run away from home, but he usually returns within a matter of days. His attitude is one of generalized rebellion against all the standards of the family.

It is a normal adolescent trait, of course, to rebel against familial standards and values. One often sees in a normal and closely knit home a child taking views which are opposite to those of his parents' political, sexual and behavioral views. Once again, it is the intensity and the frequency of the rebellion against standards which differentiates the drug abuser from the normal adolescent. The abuser becomes intensely involved with matters that are diametrically opposed to the parents' values. In a staunchly conservative home parents find that their child will work night and day for an idealistic notion of peace. A devout Catholic or Orthodox Jewish family might find a religious convert or an atheist among their children who has gone beyond normal adolescent rebellion. The children openly rebel against the sexual standards their parents have set. They are opposed to work, to school, and to "materialism." It is in some of these latter standards that the child begins to show the first signs of self-destructive behavior. Sexual promiscuity, truancy and apathy in school lead to a denial of the necessity of work and to a denial of the child's worth to himself.

It is not unusual for the child to find a way to inform his parents about this behavior. Contraceptive pills will be casually left in open view. In extreme cases a daughter will claim she is pregnant to announce her sexual activities. Telephone conversations are held in situations where the parents cannot help but overhearing. One young lady was reported to us to be discussing her plans for an abortion in loud tones, knowing full well that her parents were in the next room. The need for the parents to know stems, in part, from the need to rebel against them. If the parent remains ignorant of the child's misbehavior then it is no longer a rebellion, and it no longer produces the desired effect.

The child may rebel against parental values in a materialistic system. Especially in middle class families the child looks with disdain on "possessions." Styles of clothing, the squandering and disregard of money, and the frequency of communal living all are manifestations of rebellion against this value. It cannot be emphasized too strongly that these same characteristics will appear in a healthy adolescent. Every adolescent who walks out of the house in dirty jeans and in an old army jacket, with his hair flowing down his back, is not on that account a drug abuser. The intensity of the need for these extrinsic manifestations, however, cannot be overlooked.

Other rebellious behavior can frequently be seen as an awareness of social concerns. The child can use social progress as a rationalization for his rebellious behavior and justify it so thoroughly that it is hard to call it a symptom. It is the *motivation* for the behavior that makes it symptomatic, not necessarily the act itself. A young boy or girl, growing up in a white, middle-class neighborhood, may suddenly take on the cause of the black ghettos. He or she may become involved emotionally and sexually with a member of another race or with a racial political group, such as the Black Panthers, the Black Muslims, or the Jewish Defense League. It is not the purpose of this book to make value judgments on these various organizations, but we must understand that it is not necessarily out of purity of motives that these groups are joined.

A case history can make the point more effectively. N was seventeen at the time that he entered treatment. He was an honor student in high school through his junior year, the captain of his basketball team, and he had aspirations to become a criminal lawyer. It was toward the end of his junior year that he became involved with the Black Panthers and with various antiwar organizations. He also began to try drugs at this time. He became politically active and was recognized as an important member of the New Democratic Coalition in the area. Yet he had all but dropped out of school, as his marks slipped to subpar levels. He had nearly forgotten about his once favorite sport of basketball and he was no longer interested in "school" loyalties, feeling that they were a useless commodity in our era of racial and political turmoil.

After entering therapy, it became apparent to him that his primary motive was not that he was objectively concerned with racial or political ethics, but rather that he knew that his parents would be highly distressed by these activities. He could never consciously recognize his anger at his parents at that time, so he had to justify his rebellious behavior by placing it under the umbrella of a good cause. Following termination of therapy he did not give up seeking political and social solutions to the problems of the day, but he began to seek them in a way that was less destructive to himself. He was able to regain what he had lost in school and enter college in good standing. He stopped smoking marijuana as a practical matter; he stopped getting beaten over the head by the police for breaking the law. He developed a new relationship with his parents, not based upon the fantasy that he had entertained previously, but on the reality of the way his parents were. For the first time he could accept them as people and he did not have to rebel in anger because they were not what he wished them to be.

People whose rebellious behavior appears productive rarely seek treatment. It is not until rebellion takes the form of self-destruction, and where the self-destruction reaches a degree where it is no longer tolerable to either the individual or the parents, that they will reach out for help.

Secondary to the rebellion against the parents is the rebellion against society. Most frequently, overt action occurs in schools. The child becomes apathetic, refuses to work, cuts class frequently or misses school completely. Often the parents are unaware of his behavior; the child will manage to obtain and destroy warnings and truancy notes. At least one of these truancy notes, however, will slip through to draw the parents' attention to what the child is doing. Education, learning and marks have all become meaningless to him. Despite the reality of the need for grades for further education, the child rationalizes his lack of interest in academic success by stating that it is a "stupid principle to judge a person by his marks and this should be changed." This rationalization, based upon total denial and withdrawal from all the standards of society, contains a germ of truth. Yet the child is refusing to accept what *is* and is trying to adjust to what is *not*. Unfortunately, schooling in our culture has become so important that because of these early mistakes the youngster can seriously undermine his goals and career. An error that can affect the rest of his life, this casual attitude toward truancy severely restricts his freedom of action and hence his later alternatives.

Similar to the rebellion against the school is the child's rebellion against traditional religions, which have been accepted and taught by the parents. The child may rebel in any one of a number of ways, most frequently by not going to church. No matter what religion the parents profess, the child avoids the trappings and the formalization of that religion. Many youngsters state they have given up God, and profess atheism or agnosticism. They reject the idea that there is a just and merciful God, typically on the grounds that any such God would not allow wars, poverty and unhappiness to exist. Logical though this religious conclusion may be to some, it is often because of the need to strike back at parents that it is espoused by youngsters. In short, it is simply a rebellion.

Rebellious behavior may contain sincere motivations, of course. Young people in increasing numbers have converted from their parents' faith to other religions. The trend in the beliefs of the modern-day youngster is toward mysticism. There is a strong revival in astrology, in Eastern cults, and in the mystical aspects of Christianity and Judaism. The "nonrational" religions offer a particularly satisfying alternative to those who have been rejected by their parents on "rational" grounds. This phenomenon is not dissimilar from the turning to drugs as a magical release from anxiety; it is almost an acting out of the search for nirvana. There seems to be little "devotion" in such religious experience, as there is no devotion of time or effort to most things in rebellious youngsters' lives. They change religions quickly and facilely.

Favorite escapes of teenagers and young adults in recent times are the eastern religions that are less mystical and more practical. Some of the Indian religions, in fact, demand strict guidelines for personal living. They demand their devotees to remain celibate, refrain from eating meat, and meditate for a certain period of time each day. It seems strange that the very youngster who has completely divested himself of all the guidelines provided by his parents should cling so strongly to the tenets of a religion that are far stricter. This phenomenon becomes easier to understand when one recognizes that many of these youngsters are asking for a parental figure to exert some form of authority and to set limitations. Many drop out of school and work to devote more time to their gurus. This is usually contrary to the preachings of their teachers, but fulfills their need to live in the stress-reducing environment of a substitute family.

One sixteen-year-old girl, in our experience, from relatively orthodox Jewish parents, overthrew all of her parents' values and ran away from home on a number of occasions, with the last episode ending in her arrest. She had been highly promiscuous and had contracted a venereal disease. On return to her home she was brought to treatment by her parents much against her will. Approximately six months after entering treatment she went to a meeting held by a widely advertised guru. She was so enthralled by his teaching that she became an ardent devotee. She dedicated her whole life to the teachings of this young man and the care of his Ashram (a communal living establishment). Her parents were totally distressed by her new behavior and tried desperately to return her to their way of life. She finally consented to enter college and began her studies at the beginning of the next year. Finding the life in college too stressful for her, she made an attempt to return to her guru. At the request of her parents, however, he convinced her to return to school as she would best serve his teachings in this manner.

This young woman, like so many others, was seeking limitation. At the same time, she did not feel that she could ever accede to her parents' demands. College, vocation, marriage, and family just seemed too remote to her to ever reach. The guru demanded nothing more than meditation and prayer. The work he suggested was simple and consisted of such mundane tasks as cleaning the Ashram, begging for food, or working in a menial job that required no mental exertion. This is a far less threatening situation for the young people than are the rigors of the society in which we live. Most have such marked feelings of insecurity that they do not believe that they can ever achieve the objectives that have been set. As is the case with the aforementioned girl, despite a high level of interest, the ego functioning can be so limited that it makes the intellect practically useless.

Recently many of the drug-abusing young people have been turning to Christianity. Known by the slang expression "Jesus Freaks," they devote their whole life to Jesus and make an attempt to emulate him. It is a religious replacement for the drugs to serve the need of the individual. Religion fulfills many of the same feelings of emptiness and lack of family that had been fulfilled by the drugs. Once again the child has found a replacement for the discarded family with a benevolent, parental figure to set the rules and the guidelines.

Perhaps the most distressing form of rebellion to the parents is the sexual acting out that is so common among the drug abusers. It appears on the surface as if there is a total upheaval in sexual mores amongst the younger generation. Promiscuity is prevalent among young girls, and the girl who is a virgin at sixteen is the exception rather than the rule. Sex is not limited to one sexual partner, but more frequently there are numerous affairs. Girls will often engage in sex with boys for the drugs that they supply, but any suggestion that this is prostituting themselves for drugs appalls them. Superficially they feel no guilt and state that virginity is unnecessary. They regard the thought of the dual standard, permitting sex for boys and not for girls, as ridiculous and demeaning. Group sex has become common. Girls frequently sleep with three or four boys, or three or four couples may swap partners in a sexual orgy. Most of the youngsters will admit candidly that they often feel that they regret engaging in sex, but they find it hard to change and remain in the group. Virginity has become an object of ridicule to such a degree that many girls actually profess that they have had numerous affairs when in fact they are still virgins.

It appears, on the surface, that we have a group of youngsters whose entire way of life, thoughts, standards and mores differ entirely from the previous generation's. On first glance these are amoral, irreverent, rebellious and even criminal children, and they have caused many parents to wonder how they could have raised somebody so different from themselves. However, once you look beneath the surface, you find an entirely different personality.

Rather than being without a conscience, the youngsters really have a very primitive conscience or superego. This superego has never fully developed into that of a mature adult. Contrary to being free of guilt, the child internally is suffering from extreme guilt, because the infantile superego produces a constant dread of punishment. It is a highly punitive superego with lacuna that is governing the behavior of the child. Why then does he behave in this manner? It becomes necessary to deny the punitive superego, to deny to himself that it exists. If he did not do so, the child would be unable to function without constantly anticipating punishment. This punishment is usually either felt to be a total abandonment by the parents, or even death. With a mature superego people respond less from fear of punishment and more from the recognition of what is right and what is wrong. As one matures, he no longer is as concerned with pleasing the parents as he is concerned with doing what he knows to be right. The individual with the immature superego is constantly attempting to please the internalized parental figure in order to avoid rejection and reprisal.

The denial of the existence of this punitive superego causes a character disorder. The child rejects the entire concept of guilt, and proceeds to prove to himself that he does not feel guilty. For example, if a girl is constantly involved in sexual promiscuity, she continually reassures herself that this is how she wants to behave and that she does not feel guilt. If she stops her behavior long enough to think of what she is doing, the guilt begins to appear and the system of denial breaks down. It is here that drugs become especially useful. Sedatives suppress guilt similarly to alcohol. Inhibitions are shattered and the person behaves in a manner which otherwise would seem intolerable to him. Relationships with other people in the drug culture reinforce this denial, as they themselves profess standards similar to the child's. When the behavior stops and the child is placed in a situation to face himself, it is inevitable that he becomes severely depressed. Part of this depression stems from the guilt he experiences about his thoughts, feelings and behavior of previous years.

DEPRESSION OF THE DRUG ABUSER

Virtually every youngster who comes to a therapist for help in a drug problem has had either an overt or latent depression that existed prior to his use of drugs. Those who do not demonstrate a depression are usually suffering from psychotic, neurotic, or other character disorders. When they arrive at a treatment facility the depression isn't always manifest. It is usually covered by layers of behavioral defenses. Once the behavior is modified by the demands of treatment, however, depression becomes apparent.

It is a good question whether or not such depressions are a result of the pressures of treatment and a loss of drugs. Careful interviews with youngsters have shown, nevertheless, that the typical depression existed prior to their coming to a center and *even prior to their drug use*. Depression is often unrecognized or unlabeled by those suffering from it. It can be spotted only by the behavior one engages in as a mode of escape. Overt behavioral disorders quite commonly are defense mechanisms against the pain and suffering of depression.

Depression is an emotional illness. Unlike grief, it is not a normal reaction to loss, nor is it self-limiting, but is a conglomerate of numerous factors in the psyche. Grief, on the other hand, is a normal reaction to loss and is expressed through tears, withdrawal, and a feeling of unhappiness. Depression manifests itself in nonfunctioning behavior, withdrawal from society, feelings of resentment towards the world, self-loathing, and loss of self-worth.

There are physiological manifestations that sometimes accompany the emotional difficulties. Quite common is a disturbance of one's sleep pattern, in which the individual suffers from insomnia or excessive sleep of twelve, fourteen or eighteen hours a day. Likewise there is often a loss of appetite and weight and resulting malnutrition, or compulsive eating and obesity. Bowel changes are a common accompaniment of depression; generally there will be constipation, but it can be manifested in unusual frequency or diarrhea. There may be an overall sense of body fatigue and loginess, with a lack of motivation towards either a productive goal or towards pleasure.

The absence of pleasure in depression takes the form of an anhedonic existence, in which the depressed person actually revels in unhappiness. Thoughts of suicide are frequent, and with greater severity of depression the chances of a suicide attempt are increased. This suicidal desire is often substituted in the drug abuser by self-destructive behavior that serves the same ends. Sometimes the need for self-destruction is to punish an unrealistic or irrational guilt that is one of the cornerstones of the depression. Such guilt is so overwhelming that an individual is rarely able to cope with it except through some form of punishment.

The cause of depression, according to Freudian theory, is based upon object loss. The object need not be a person, but may be a materialistic object, an idea, self-esteem, or anything for which the individual has a strong libidinal attachment. Also, the loss need not be actual, but can be simply felt or imagined. The loss of a loved one, money, a job, one's hair or one's self-esteem can have equally devastating effects on an individual if enough libidinal energy has been attached to that object.

A second criterion of depression, along with the loss of an object, is the need for the individual to have introjected the loved object. Introjection is a psychic method of incorporating the loved object into one's self. A primitive mechanism, it usually takes place in the oral stage of a child's development. It is part of the natural growth process of the child, the first step in identifying with one's parents or one's heroes in order to develop an ego ideal, a sense of self, and a superego. It is seen in hero worship among youngsters, or the desire to be exactly like the parent. As we have pointed out earlier, *incorporation* is different from *identification* in that it is a mechanism which does not differentiate parts of an individual but consumes him or her whole; in identification, there is a selective "digestion" of the individual and the elimination of those aspects deemed undesirable. The incorporated object is taken in wholly without the exclusion of any of the negative characteristics of that individual. During the process of maturation, a person begins to identify with the same love object, and he begins to excrete from himself those characteristics which he would not want as part of his personality. Identification is obviously a more mature mechanism than is incorporation.

When the love object has been introjected at an early age in the person's psychic development, many unrealistic expectations from that object remain. When those expectations cannot be realized the individual becomes enraged at the object. Because of the intensity of the rage and the fear of losing the love object, the rage has to be suppressed and ambivalence develops. A youngster may have the expectation that his parents will always be present for his needs. When his parents cannot live up to this expectation—either because of other needs, other pursuits, other children or even death—then the youngster develops intense anger against the parents.

A twenty-one-year old man lost his car in an automobile accident. Though he escaped unhurt, he went into an intense depression at the loss of the car. He had invested the car with the symbols of his manhood and felt that he was devoid of any other masculine traits. When the car was demolished he became so enraged that he set fire to the automobile. He then withdrew to his room and eventually returned to the use of heroin. His depression stemmed from a rage at the world and at "fate" for having deprived him of his beloved automobile. The rage then manifested itself towards his parents, whom he magically blamed for the loss of the car. "They never wanted me to have it in the first place," he stated during a session.

His belief in his parents' omnipotence directed the rage from the world to the parents. Totally dependent upon his parents and unable to express his rage, he became exceedingly suicidal and self-destructive. He was, in essence, turning the rage against the *introjected* parents rather than at the actual parents. This mechanism of turning the anger against oneself is a crucial step in a depression. However, the anger is not psychically directed at the whole self, but rather at the part of the self that is made up of the introjected object. A suicidal knife, plunged into the body, is not directed at the body which takes the stab, but at some object dwelling within the body in the form of the introject. In short, turning of anger from the outside object towards the self creates a depression.

There are many secondary effects of a depression that work in a vicious circle: as the depression increases, the symptom increases, which causes a further increase in the depression. One of the most difficult hurdles to overcome in the treatment of depression is the removal of the secondary "gain" which the individual feels, of making everyone around him miserable. An unconscious gratification of the depressive is knowing that the people around him are suffering.

The secondary gain of punishment of other people is further heightened by the self-destructive and suicidal nature of the depressed individual. Most significant in the drug abuser is *the self-destructive aspect of the chosen punishment*. He entertains an unconscious fantasy that he will be causing discomfort, guilt and unhappiness to those people at whom his rage is directed. There is little difference between the motivation of the suicidal person and the self-destructive person. The secondary gain of both is the fantasy that pain will be suffered by the individuals at whom the anger is initially directed.

A depressed person feels an accumulated sense of frustration and helplessness about being unable to reach the people at whom his anger is directed. The suicidal individual feels that these people are unresponsive to his pleas and supplications and cannot be reached by ordinary methods. Usually there is an unconscious belief, as a remnant from infantile thought, that these parental figures have the omnipotent power to know and to understand his feelings without need for verbal communication. There is so much difficulty in the verbal expression of the anger that the person never does express it directly and the recipients of all this rage are usually ignorant of how they are causing the depressed person to suffer.

When the fantasies of the suicidal individual are delved into, a consistent pattern emerges. He may express directly that his fantasy includes the people at whom he is angry, usually his parents, brothers, sisters, wives or loved ones. He mourns their fate and cries over their loss. The suicidal person expects that others will experience a feeling of guilt, regretting the suicide and wanting to bring him back to life. But it is usually only a wish; he feels that they will accept his passing with the same lack of caring he feels from them while he is alive. Unfortunately, this is often true. The people who are the object of suicidal aggression are generally those who are not responsive to the individual's pleas. It is for this reason that suicide gestures are so frequently repeated with increasing violence each time. This increase in the intensity of the acts is equally true for the self-destructive person.

A typically self-destructive youngster at the age of thirteen ran away for two days without getting the desired response from the parents on her return. Two weeks later she ran away for a week. After she returned this time the parents were concerned, but she soon began to feel her control over them slipping. After a number of months she once again ran away, and this time returned home pregnant. Unfortunately, once again the parents were not responsive to her infantile needs, but treated her in an increasingly remote manner as their own anger towards her grew. As she sensed this widening gap she became more and more desperate. Following an abortion she left home again, this time to stay away a number of months. On her return she was totally debilitated and addicted to heroin. The only result of this behavior was to have her parents bring her to court, where she was put on probation and sent to a treatment center. It was not until after she had left the therapeutic community and entered psychotherapy that she began to understand the motivation for all her self-destructive behavior.

Her unconscious fantasy had always been that her parents preferred a brother, who was two years younger. She felt that she had been deprived of her mother. When she entered adolescence and faced the conflict of dependence versus independence, there was a renewed desire for parental nurturing. There was a good deal of reality in her sense of loss. Her mother was inadequate in coping with two children. There was some negligence in that her grandmother had been responsible for most of her early raising. When her brother arrived, her mother concentrated most of her feelings on the younger child, depriving her daughter.

It is unlikely that such a severe reaction to the loss of nurturing would have taken place strictly as a fantasized loss. The depression that began in early childhood because of an actual loss became intensified throughout the years, culminating in an unconscious depression in early adolescence. When she entered treatment, there was no awareness of the depression which she was suffering, but as therapy progressed, early memories of a very unhappy childhood, with hours of loneliness in her room and with long spells of crying, came back to consciousness. Her self-destructive behavior was motivated by the desire to regain the parental nurturing that was so badly missed and by the desire to punish the parents for deserting her in her childhood. Of course, the fantasized punishment was based upon the feeling that the parents would become increasingly grief-stricken and guilty as they watched their daughter slowly destroy herself. The wish, of course, was never fulfilled.

The desire to punish, it should be emphasized, is always directed at the parent or the surrogate parent. Even when the person against whom the rage is ostensibly directed is one who arrives in later life, he is inevitably a representation of an early parental figure. So frequently do adult relationships recapitulate the same traits that were found undesirable in the parent that we cannot help but conclude there is a strong need to repeat negative childhood experiences. Freud called this pattern a repetition compulsion. It is based on the desire to bring about the expected change in a later person that the infant or child was unable to bring about in the parent. It is an impossibility, of course, to repeat the feelings of helplessness and frustration that were found in childhood. The resultant anger is then directed at the parent substitute *with the resultant desire to punish him or her through the punishment of the self.*

A combination of factors occurs in depression which makes it a self-perpetuating system, not requiring the initial stimulus maintaining intensity. In the pre-depressive personality there is inevitably a primitive superego structure. It is highly punitive, based upon the fear of parental displeasure and severe guilt, with the desire for a deserved punishment. It does not permit of any unacceptable feelings or thoughts, and it fails to differentiate between thoughts and actions. The depressed individual can feel as guilty for thinking of a violent act as he can for committing one; the anger that he feels toward the parental figure, therefore, becomes a source of self-condemnation. He is guilt-ridden because of the destructive impulses that he has towards the parent substitute. He is fearful of parental retribution in the form of abandonment for his unacceptable feelings. Furthermore, many of his actions are unacceptable to himself as they are in opposition to his ego ideal. As a result of his desire to punish others he himself has become a nonfunctioning, nonacceptable person.

He begins to lose self-esteem, which eventually turns to self-loathing. He no longer feels that he is capable of functioning and no longer believes that he deserves positive accomplishments. Success is now unacceptable because of his desire for the superego to punish the unacceptable part of the self. The lack of success increases the feelings of inadequacy that are already present, which brings about further nonfunctioning, with a further decrease of self-esteem, and a further increase of self-loathing. We now have a self-perpetuating cycle, though the initial cause of the depression has been long repressed. What is now apparent is the sense of inadequacy, the lack of self-esteem, and the intense self-loathing. As this continues, the person eventually must totally withdraw from a functioning society. He will often spend most of his hours in his house or in his room, neither working with nor socializing with others.

In the formation of the primitive superego, the parental image has been incorporated and parental standards have been accepted in toto. Many of the individual's antisocial acts are manifested in a behavioral disorder, which acts as a defense against depressed feelings, and is a source of further guilt. The sexual acting out that so commonly accompanies adolescent behavioral disorders is still unacceptable to the youngster's superego. Truancy, stealing, and drug-taking will all increase the guilt and perpetuate the unconscious depression.

Depression in a drug-abusing individual seems to follow fairly consistent patterns. As we have seen, intense oral needs of the infant have never been met because of either a physically or emotionally absent parent. The lack of gratification of oral needs persists through the individual's life, with the resultant consistent craving for the gratification of these needs. It is worth repeating that even in those individuals who manifest an independent appearance, there is usually an intense dependency need. The drug-user's independent behavior is a facade. Because of his unfulfilled oral need, the individual remains in a state of rage towards the parental figure. This underlying rage has to be repressed because of the fear of destroying the individual on whom he is so dependent, accompanied by the fear of abandonment by that individual as a punishment for the angry impulses of the youngster. The rage is then redirected against the incorporated parent and subsequently against the self as a defense.

In the adolescent years there is a resurgence of oral needs. The child becomes increasingly fearful of his role of true independence, as he breaks away from the family, turns to the outside world, and senses his upcoming adult responsibilities. Knowing that these dependency needs cannot be met, the adolescent manifests increasing anxiety with resultant anger towards the people who fail to meet his oral needs.

Two paths now develop that alternately converge and diverge. First, there is the self-destructive urge against the incorporated parent, with the subsequent desire to destroy the representation of the absent parent. This leads to much of the self-destructive behavior that is seen in the adolescent drug abuser. Second, there is usually turmoil in the parents, to whom his anger is primarily directed. His parents are in a state of anxiety, bewilderment, and agitation that is unconsciously gratifying to the young drug abuser. Because of his primitive superego, however, the youngster feels intense guilt and fear of his parents' rejection. He therefore finds ways to overcome his guilt and to hide his depression.

His primary defense against guilt is a denial of it by means of an increasing degree of antisocial behavior to prove, subsequently, that his behavior is really "egosyntonic," in concert with his true self. For all intents and purposes his behavior actually does become egosyntonic and the child finds no reason to give it up. The depression does not become manifest until the behavior is stopped, and so it is necessary for the child to fight against any attempt to end the antisocial behavior.

Despite the outward appearance of confidence and rejection of all values, the depressive is still the basic, very highly dependent adolescent. His dependency is far more intense than in a normal adolescent, who manifests his dependent wishes and actions more openly than does the drug abuser. Fear and anxiety, as we have seen, turns the drug user from any productive behavior. The stereotype picture of the drop-out emerges. Yet the child cannot basically accept this within himself, and beneath a rather thin veneer he becomes filled with self doubt. Lack of functioning has diminished self-esteem to a point where he feels totally inadequate.

When such a youngster is forced into situations where he must function, the terror becomes so pronounced that he seeks constant reassurance that what he is doing is acceptable. Most of the youngsters who enter the final stage of treatment in a drug center are in such a situation. They are required to find work and to socialize with people who have not been in drug centers. The number of telephone calls received by the center each day from these youngsters seeking reassurance and support is phenomenal. Despite their high intelligence and musical and artistic talents often far superior to the average, they are convinced that they are incapable of competing in society. It is the fear of total annihilation of an infant abandoned and alone. This lack of self-esteem and the overwhelming feeling of inadequacy leads to a sensation of hopelessness that becomes a major factor in depression.

The two paths once again converge as the child who feels so hopeless and inadequate becomes intensely desirous of a parent or a parent surrogate. Because there is no person who is able to fill the role and to allow him to return to the infantile state, the intense rage, once again, begins to emerge. As the rage rises against the dependency object, the individual must again repress it and again it turns against the self.

It is at this point that, in treatment, the youngster frequently reverts to the pattern that we have seen previously: the behavior again becomes self-destructive. He no longer finds himself capable of escaping into the world of drugs because the superego has incorporated the values of the drug center. His self-destructive behavior becomes manifest in far more direct means. There is an increased frequency of suicide gestures, of running away, and of repetitive failures. Because of the availability of understanding people with prior work in treatment, the drug abuser is now able to express verbally the fear and depression that he is encountering. His rage itself becomes manifest, and in fantasies and dreams it begins to become externalized. Those individuals who do not have the benefit of a therapeutic situation, however, revert to drug use the moment that they are forced into a productive situation. A few case histories will help show how this theoretical framework can be quite a practical aid in therapy.

J was nineteen years old when he entered treatment. He was above average in intelligence and was a graduate from one of the better prep schools in New England. He had been accepted by a number of colleges, one of which has a reputation for the highest standards. While he was in prep school he began to abuse many of the softer drugs, until, shortly after graduation, he turned to heroin. He was arrested numerous times on drug charges, for burglary, and for felonious assault. It was because of the pressures of the court that he entered therapy.

It came out that he was adopted by his present parents when he was two-and-a-half years old, having lived in a foster home from the age of eight months. He had been told of his adoption in his early years, on the advice of the adoption agency. His new parents were concerned and loving, though they found it difficult to express physical affection. In any other situation there is a strong possibility that the child would have grown to be relatively free from severe neurotic traits, but the adoption presented problems.

J knew some things about his natural mother and had fantasized many more. Throughout his life there had been a constant desire to seek her out and to find out more about her. He knew that she was a girl of fifteen who had become pregnant by a man in his forties. Furthermore, he knew that this was the only relationship that she had had and that she was not a "wanton" woman. He also knew that he was placed for adoption because of the mother's immaturity and inability to care for a child. His early memories of his foster parents were that they were warm, loving people, who unfortunately could not keep him past his second year. He vividly recalls the day that his adoptive parents came to pick him up and how, despite the fact that they brought with them toys and games and personal warmth, he always wanted to return to his foster home. That memory was so emotional that he could not relate it without crying. He learned to love and accept his new parents; yet at the back of his mind there was a constant, nagging fear that they too would eventually abandon him.

The fantasy remained from childhood that he was in some way responsible for having to be given up for adoption. It made him feel worthless and useless as an individual. When he had finished prep school and became aware of the intense responsibility of college life, his first severe anxiety began. This anxiety reappeared every time he was placed into a situation of productivity. He never felt equal to the tasks at hand, suffering from overwhelming feelings of inferiority that stemmed from his early days of childhood and the beliefs of his own worthlessness. He became increasingly depressed from the age of fifteen on and eventually found relief from depression in the use of drugs.

His interpersonal relationships were inevitably poor and he maintained strong hostility towards women. Any young girl became a displacement of his natural mother, whom he still resented for giving him up for adoption. He further found himself in difficulty with older men with whom he associated, as he displaced the anger he felt for his natural father to intense feelings of loathing towards them.

He was finally forced into treatment as a condition of probation. In treatment, a relationship developed rather quickly, as he was so hungry for a sustained relationship of any kind. He was able, through the use of this new association, to begin to veer away from drug abuse and eventually gave it up except for marijuana. Throughout his entire treatment he was never able to give up marijuana because of the tranquilizing and antidepressant effects that it had on him. He maintained himself, free of other drugs, for over a year, but throughout that time he remained nonproductive.

Finally, after a year, stress was placed upon him to begin to search for a productive way of life. He took menial jobs in which he felt confident he was able to succeed, but, in a short time, he found that they further demeaned his already limited self-esteem. He became angry and distressed at the work situation and he vented his anger on his boss, leaving the job within a period of weeks. At no point was he able to maintain a job for longer than a month and a half. The jobs were always either in a factory or in a dog kennel. Any attempt to get him to assume a greater responsibility was always met with the fear and anxiety that he could not succeed at any demanding task. He was caught, therefore, between Scylla and Charybdis: on one hand, the feelings of inferiority which did not permit him to accept a task from which he could receive gratification; and on the other hand, the shrinking of self-esteem and the feelings of worthlessness that were caused by menial work. Furthermore, whenever he took a job, within a period of days he began taking Quaaludes or barbiturates in order to alleviate his anxiety. An attempt was made to treat him with phenothyazines and once with antidepressants, with little success. The medication had little or no effect on his constant depression.

Though this seems to be an extreme case, the depression mechanism at work is typical of many youngsters who have an overwhelming sense of inferiority and worthlessness. It seems impossible for them to face the anxiety of productivity, and the feelings of worthlessness that develop enhance their depression and further complicate the task of changing their life style. The need to avoid productive pursuits is termed a "secondary" reinforcement of the self-destructive urges that are always present. By a system of rationalization and denial, the drug abuser tries to relieve his feelings of worthlessness that are caused by his style of life. A further complication is the guilt caused by not adhering to parental values, which have never been completely denied by the youngster.

In girls, increased feelings of worthlessness and lowering of self-esteem are usually brought about by frequent sexual activity that accompanies the drug abuser's life style. Most of these young girls find themselves in sexually compromising positions because of feelings similar to those that discourage productivity on the part of boys. They become easily swayed by others because of the feeling that they are not desirable in themselves, but can only become desirable because of their sexuality. They indulge in all varieties of sexual practices to obtain a boyfriend or to hold onto one, and frequently prostitute themselves to obtain drugs. Despite the facade they exhibit of their sexual liberation, these girls retain value systems similar to those of their parents; invariably, they develop serious guilt feelings.

R was seventeen when she entered the drug center. She was referred by her parents through pressures placed upon her by the court after repeated drug abuse. In the neighborhood she was sexually promiscuous, with a reputation for being an "easy mark" for any boy who wanted a girl for the evening. She had been using downs, which she found alleviated the sense of guilt and self-anger which pressed upon her because of her sexual practices. When she first entered the center she was accordingly able to maintain a denial of guilt.

She was the younger of two daughters, in a family that placed strong emphasis on "maleness." Her parents had wanted a son very badly, and on her fell the responsibility of being the "male" child of the family. She learned at an early age that in order to gain acceptance from her father she would have to pursue a boy's activities. Her favorite pastimes soon became softball and basketball. She spent many hours with her father playing catch. As she grew up she began to develop a markedly confused identity, and with the onset of menstruation she began having severe troubles surrounding her periods. She never felt accepted for being a girl, and was always uneasy about her parents' love. Because of her strong attachment to her father she became increasingly fearful that this attachment had sexual overtones.

In early adolescence there was a sudden and complete break from her father and her mother. She turned to the streets and to drugs. Though she is an attractive girl—intelligent, personable and easily liked by all members of the treatment center—at that time she was sure that she could not be accepted by men as a woman. She resented her femininity and anything that reminded her of it. She developed strong, unconscious, hostile wishes towards all of the men that she met, and soon became a sexual object to them, often using her sexual relations to act out her hostility to them. Her fantasy was that if she were desired strongly enough for her sex, she would gain a control over men. She would then not run the risk of losing them, while at the same time she would have active fantasies of tantalizing, seducing and eventually frustrating them. She was, in essence, playing out with men of her age hostile feelings which she had repressed towards her father for not accepting her for what she was.

Not long after entering therapy she began to discover an overwhelming guilt that had been repressed about her sexual behavior. She became increasingly depressed with strong feelings of worthlessness. She gave up all sexual contact with men; the only contacts she maintained always had hostile overtones to them. Her depression became increasingly intense as the guilt became more conscious to her. Much of the rage that she felt towards her father was also turned upon herself; in the ambivalence of her feelings she could not tolerate the thought of losing him.

This is a clear example of the combination of inwardly directed rage and of loss of self-esteem—through the self-destructive behavior that increases the depression—in many young girls who have turned to drugs. Once the initial barrier has been broken through, depression over the sexual acting out becomes apparent, and most of the sexual behavior comes to a stop. In fact, sex becomes excessively taboo.

Boys suffer too from the demands of their sexuality. They also suffer from marked fear of inferiority and very often turn to sex to prove their masculinity. Sexual prowess among younger adolescents is regarded as a mark of manhood. Most of these youngsters, however, are more fearful of sex than they are ready for it. They find themselves forced into situations of marked anxiety because of the pressure to prove their own competence. Because the anxiety level is so high, their sexual performance is usually of a very poor quality, and this fact further discourages them about their masculine images. As they become increasingly uncomfortable, they turn to marijuana or barbiturates to cover their anxiety. Though sexual prowess is often increased with the use of marijuana, it is markedly diminished under the influence of barbiturates; use of the latter further increases their feelings of sexual inadequacy. As they feel less adequate they become more depressed; to mask depression and feelings of inadequacy they more actively turn to drugs, and a vicious cycle mires them deeper in self-doubt.

Surprisingly, at this point it is not unusual for many youngsters, both boys and girls, to turn to homosexuality. They attempt to reach out to some fellow human being in order to establish contact. Their homosexuality further increases their anxiety. They question further their own masculine or feminine image.

With homosexuality there arise strong projective defensive mechanisms that often lead to blatant paranoia. The incipient paranoid state is diminished with the use of barbiturates, but when barbiturates are discontinued there is a marked resurgence of the paranoia. On the other hand, marijuana enhances paranoid feelings, and so youngsters increasingly choose barbiturates, with further harm to their sexual self-esteem. The paranoid feelings in these cases are less a symptom of schizophrenia than they are of feelings of worthlessness and severe depression. A new mechanism takes hold: the anticipation that others are going to view them in the same way that they view themselves.

Anticipation of peer disapproval leads to further behavior designed, unconsciously, to bring about the very thing that they fear the most: rejection by others. The anticipated rejection stems from the displacement of their feelings about their parents onto the group. As primary objects, parents are the ones whom the individual fears losing the most. Because of an unconscious awareness that so much of their behavior is displeasing to parental ideals, they anticipate that they will be rejected by any parental figures. This supposed rejection is coupled with a projection of their own feelings of worthlessness and low self-esteem, adding to the anticipation that others will see them in a negative light. The *actual* rejection they experience from others is carried out through many behavioral techniques, specifically designed by the unconscious to bring about signs of rejection and to prove to them that their assumptions about themselves are right.

Once alienation from their peers has taken place, there is a further decrease in the gratification of their oral needs. They are led to further anger and to the further need for self-destructive behavior. It is apparent that at this point their depression is self-perpetuating, and it is precisely here that drugs begin to make their inroad. Drugs become one of the few ways of breaking the chain of depression, guilt, and alienation.

—9—
The Drugs Children Use

Physically addicting drugs are the most dangerous to children. The physically addictive drug causes the body to develop a physical dependence, and when the drug is withdrawn there are actual cellular changes that cause severe physical symptoms. The two most common addictive drugs are the narcotics and the central nervous system depressants—often classed under the main subcategory, barbiturates. Though usually not included in this group, it is safe to consider alcohol as physically addictive; the degree of the addictive potential of alcohol is, however, below that of the others.

A second category is that of the psychologically addictive drug, which, for the sake of clarity, is referred to as habituating. In this category falls the vast majority of the drugs that are the most commonly abused. These are the drugs which cause the individual to develop a dependence: drugs become a necessary mainstay for the individual's psychological sense of well-being. They diminish his anxiety, increase his ability to cope with stressful situations, and allow for a dulling of the emotions that cause him pain. Upon withdrawal of this type of drug, there is no actual intracellular change. Yet the individual does exhibit physical signs and symptoms, which are related to those which accompany severe anxiety. There is extreme nervousness, tension, sweating palms, a sense of discomfort and anxiety. The withdrawal is not nearly as dangerous as that of physically addicting drugs. But it is often equally as *difficult*. Included in this category of drugs are the stimulants, the hallucinogens, tranquilizers and pain killers, and the numerous drug-substitutes: glue, ether, nitrous oxide, etc.

Marijuana and hashish are categorized under the hallucinogenic group. There has been no scientifically substantiated evidence to date to indicate that cannabis, used in moderation, has physically deleterious effects. Therefore, not all users of these drugs can be called "drug abusers." On the other hand, much like alcohol, this drug type is frequently abused to an extent where the individual loses his ability to function and becomes psychologically habituated to the drug.

Nicotine, too, falls under the category of a psychologically habituating drug, as anyone knows who has made an attempt to give up smoking. It is not considered here except as a "signal" from a parent—because of the fact that it does not alter mood or behavior to the extent found with other drugs. I do not, however, wish to imply that this makes it less dangerous to the user than many drugs. Its sustained use is patently toxic.

The narcotic drugs include opium and its derivatives, heroin, morphine and codeine; synthetics, such as demerol and percodan, and most recently, methadone. The basic desirable effect of the narcotic drugs is to produce a sense of euphoria, accompanied by an emotional apathy. The concurrent reduction of anxiety and tension produces what has been described by some as a state of near bliss.

The narcotics are taken into the body by one of three routes. The most desirable method for the addict is to "mainline"—to take the drug intravenously. This method of induction causes an immediate sense of warmth, commonly known as "a rush." It is this experience that is described by some addicts as the most desirable part of the drug experience. Following the initial rush, the addict experiences a sense of relief and relaxation with an eventual drowsiness and dozing off to sleep, commonly known as "going on the nod." It is this sense of relief from the cares and pressures of the outside world which is what psychologically maintains the addict in his desire for the drug.

Heroin and morphine are the most commonly used drugs for injection; increasingly, methadone has been taken intravenously. The second most common route, recently, has been oral. This has become increasingly popular since the advent of methadone, which is basically taken by mouth. Methadone and codeine are also orally ingested. Though they do not produce the same rush or the intensity of the "high," which is a state of euphoria, they do produce a sense of relaxation, or calm, and a reduction of the basic bodily drives. The third route of ingestion is nasally, common with heroin and opium, in the form of "snorting." Though snuffing does not produce as great an initial rush or the intensity of a high, it does produce similar effects to those of drugs taken intravenously.

The psychological makeup of the drug addict is discussed elsewhere; at present, suffice it to say that the addict is generally a highly dependent individual who has a great deal of difficulty in coping with the stresses of everyday life. The desirable effects of a drug involve a superficial state of well-being. It weakens the sexual and aggressive drives of the individual, makes "striving" far less desirable, and so diminishes the feeling that attainment of goals is necessary. With the apathy which is then developed, the anxiety associated with achieving these goals is further diminished. To the addict there is nothing unpleasant about the sensation of heroin. In fact, it is the very pleasantness of the reaction to the drug that makes it so difficult to motivate an individual to give it up—and to have to face the stresses and pressures of daily life.

The primary danger of a narcotic is, of course, addiction. The physiological addiction that is produced by a drug is brought about by intercellular changes, causing the cells themselves to become dependent upon the presence of drug molecules in the system. Withdrawal symptoms that are experienced when the narcotic is not available are extremely painful. They include such visible symptoms as muscular pain, abdominal cramps, irritability, tremors and jerks in the limbs, runny nose and watery eyes. The only relief that the addict experiences from these symptoms is from further injection of the drug. The physiological craving added to the psychological dependence together make the abandonment of the addictive state highly undesirable to the addict. Despite the severity of the symptoms that accompany withdrawal, deaths have not been reported during even the most severe withdrawals. In this sense, addiction to narcotic drugs is not as severe as that which is found with the barbiturates.

Death can occur, however, with an overdose of any drug. This occurs either during an intentional suicidal attempt or by accidental overdoses, which most frequently occur when the narcotic substance is in a purer form than the addict suspects. Because of the extremely profitable market in dealing in narcotics, it has been the practice over the last thirty years to dilute what the addict receives with one of a number of substances. Most commonly used is fruit sugar or dextrose. Drugs can also be cut with quinine—a standard practice among dealers. Though it diminishes the amount of heroin that the addict receives, cutting with quinine still produces the same initial rush because of the physiological response to the quinine. Quinine is, of course, far cheaper to produce and far more easily obtained than is heroin. When an addict believes that he is using a highly diluted bag of heroin and actually has received heroin in a purer form, he will often overdose. The result can be coma, respiratory failure, and eventually death. The basic treatment for a narcotic overdose is the intravenous use of the narcotic "antagonist." Nalothan, for example, counteracts the narcotic at the nerve endings and arrests its physiological progress in the body.

The use of narcotics can be detected by various signs and symptoms. Narcotics cause the pupils of the eye to constrict, commonly known as "pinning." This symptom will appear regardless of what method is used to ingest the narcotic. For those who mainline drugs, the physical symptoms will include needle marks or scars in the side of the arm opposite the elbow, or in the legs, in the hands, or in almost any other available injection site that the addict can reach. The presence of hypodermic syringes, eye droppers, burnt spoons or bottle caps, blood stains on the shirt sleeves or the presence of glassine envelopes or tourniquets are all indicative of a narcotic addict. The more subtle signs and symptoms include an apathetic response, drowsiness, incoherent or slurred speech, slowness of breath and general lethargy. If the individual is "on the nod" he will be in a sound sleep from which it will be difficult to awaken him.

As with all addictive drugs, continued use causes an increased tolerance to the drug. This in turn requires increasing dosages of the drug to maintain the same level of euphoria. As the need for the drug increases so does the increase in cost to maintain the habit. It is the ever-increasing cost that turns so many addicts to a life of crime—for there is little else that can supply the forty, sixty, or even eighty dollars a day that many addicts need to sustain their habit. A dose of heroin that can kill a nonaddicted individual is very easily tolerated by the addict.

Hypnotics are the second general category of addictive drugs which we have described above as central nervous system depressants. They are more commonly used than are the narcotics. Hypnotics are better known as "downs," and are either barbiturates or nonbarbiturates. In the former group are included seconal, nembutal, tuanol; in the latter group are doriden and methaqualon. Methaqualon is referred to by the addict as quaaludes, quads, luds, or sopors. The common slang terms for the barbiturates are reds, yellow jackets, or frequently now, various trade names.

The barbiturate and nonbarbiturate hypnotics produce intoxication that in many ways is not too different from an alcohol intoxication. The individual experiences a relaxed sense of well-being, a drowsiness, and a view of the world through a haze. They reduce anxiety and they relieve a sense of guilt. Many youngsters find barbiturates necessary in order to soften the guilt that they feel because of their sexual behavior. When an individual ingests an excess number of hypnotics (as with alcohol), but not enough to overdose, he can become belligerent and aggressive. Eventually the drowsiness leads to a sleep state—which, in many ways, differs from normal sleep in depth and in the diminution of rapid eye movement. The lack of REM, as it is called, is indicative of a dreamless sleep and so often prefigures other minor disorders in the cycle of daily routine.

At the time of this writing, the most commonly abused of all of the downs are quaaludes. There has been a misconception among their users that quaaludes do not have the addictive quality of the barbiturates. Unfortunately, this is untrue; an addiction to quaaludes can develop as easily as it can to a barbiturate. There are two major risks in becoming addicted to downs which do not exist with narcotics. First, there is of course the danger of an overdose. Because of the central nervous system depressant effect, an overdose of downs can lead to respiratory failure, coma, and even death. Second, sudden withdrawal from the barbiturates can lead to severe convulsions and death. It is for this reason that a barbiturate addict requires hospitalization for withdrawal from the drug. It is important to monitor carefully the withdrawal from the high barbiturate levels that he attained during the increasing tolerance to the barbiturates. Quaaludes, sopors and doriden may produce similar withdrawal effects.

In case of an overdose of barbiturates, often taken intentionally in a suicide attempt, it is important to lavage the stomach of all remaining contents, to maintain respiration, even artificially, and to supply supportive measures. In the case of convulsions due to barbiturate withdrawal, the administration of barbiturates is necessary to prevent further convulsions and death.

The most common way that barbiturates are ingested is orally. We have seen, though, it is possible to inject barbiturates intravenously or subcutaneously. In the case of the subcutaneous injection and frequently in the case of intravenous injection, where the addict has missed the vein, severe abscesses become apparent.

The symptoms most commonly seen in barbiturate use are those of drowsiness, slowness of movement, slurred speech and incoherence, poor judgment and perception and eventually sleep. Hangovers are often reported the following day, much the same as with the use of alcohol.

The Federal Drug Investigating Committee of 1973 brought to light what had been suspected all along by those involved in the therapy of the drug addict. It became clear at that time that the barbiturate and the nonbarbiturate hypnotics were the most commonly abused form of drugs in the United States. Because of the vast quantity of these drugs that are used by adolescents, they constitute the major danger to the drug-oriented population exclusive of alcohol.

Though it is not a common concept, alcohol clearly belongs under the category of an addictive drug. Its potential for addiction is lower than that of the opiates or barbiturates, but because of the sheer number of drinkers, it is the most common addiction. Physiological dependency to the drug develops, and rapid withdrawal of the drug can lead to delerium tremens, the "D.T.'s." Tolerance increases, with ever-increasing amounts becoming necessary for the desired effect.

The effects of alcohol intake are well known. In many ways they are similar to other central nervous system depressants, the barbiturates. Small amounts seem to act as a stimulant, but this may merely be the depressing of certain inhibitions. A person becomes mildly euphoric, self-confidence increases, and often verbal and physical responses improve. With increasing amounts, the depressive effects of the drug become apparent. Speech becomes slurred, the gait becomes staggering, and the thought processes and powers of judgment deteriorate. As a person's tolerance to alcohol increases, it requires a greater quantity to produce these symptoms. Therefore, many alcoholics never really give the appearance of being drunk. Withdrawal from alcohol produces shakes, cramps, nausea, severe anxiety, profuse perspiration, restlessness, and finally "D.T.'s." An intense craving for alcohol is experienced.

It is probably easier to discover that a child has been drinking than to detect the use of any other drug. Not only are the symptoms easily recognized, but the odor, nearly always present, and the inability to use a small container (bottles are more easily noticed than small vials or envelopes) make detection easy. Most commonly used by today's youth is wine, often in conjunction with marijuana, followed by beer and then hard liquor.

The fact that a youngster is "only drinking" is not necessarily "better" than drug use. Alcoholism is insidious, dangerous and *legal*. It destroys the liver and brain. It damages the stomach, intestines, heart, blood vessels and kidneys. It makes drivers under its influence a menace.

Along with alcohol and the barbiturates the most common drugs of abuse are those that are not physiologically addicting, but rather psychologically habituating. Though there are no intercellular changes brought about by the use of these drugs, causing the cells to require the drug in order to function normally, there is strong psychological craving for the drug. The psychological craving, however, can almost equal in intensity the physiological craving for addictive drugs. Among the drugs in this category, perhaps the most potentially dangerous in terms of severity of habituation are the stimulants.

The stimulants, known commonly as "ups," are found in the medicine cabinets of many homes. *They are an integral part of most diet pills that are in use today*. Most commonly abused of the stimulants are dexedrine and meth-amphetamine, which fall into the amphetamine category. These drugs, commonly known as "speed" (technically, speed refers only to meth-amphetamine) can either be ingested orally, or, in the case of meth-amphetamine, injected intravenously. Cocaine, or "coke," is becoming increasingly more popular in its usage, but, because of its limited supply and its extreme expense, has natural limiting factors. Cocaine is usually ingested nasally, in the form of snuff, or intravenously.

The desirable effects of these drugs are the state of euphoria, excess energy, and physical well-being. The individual becomes more alert, more aware, and feels that he has a great command of his faculties. The amphetamine group is often used in combination with barbiturates in order to alleviate more of the unpleasant side effects, such as excessive nervousness, restlessness, anxiety, and irritability.

Like the physiologically addictive drugs, the stimulants require increasing doses of the medication in order to produce the desired effects. With the sudden withdrawal of the drug the individual goes into a "crash" which produces severe emotional depression, drowsiness, listlessness, apathy, and a total lack of energy. In order to avoid "crashing" a person may continue to take the drugs far beyond the time which he had originally intended.

The effects of cocaine are similar to those of the amphetamines, except that the period of effect is shorter and the crash can be more severe. For those people who are involved with cocaine it is "the ultimate drug." Cocaine usage is on the increase presently for those individuals who are now using methadone to replace the scarcer heroin. Many people on methadone-maintenance programs also become addicted to cocaine, as it supplements the high which methadone fails to provide.

The dangers of stimulants are numerous. The chronic user of stimulants undergoes the risk of marked appetite decline and sleep disturbances. There can be extreme weight loss, nervousness, all the fatigue situations from lack of sleep, and the marked "crash" after a period on the drug. In order to counteract these side effects, barbiturates are often taken up; the subsequent danger is barbiturate addiction. The occasional and chronic user run the risks of severe anxiety, persecutory paranoid thinking, and often exposure to total psychotic situations complete with delusional systems and auditory and visual hallucinations. Overdoses, when extreme, can lead to death. As mentioned, severe psychological dependence, almost tantamount to a physiological dependence, may occur with the drug. The use of methadrine intravenously can cause such severe appetite loss and such severe lack of sleep that the individual will suffer from malnutrition, undernourishment, or emaciation. In time, all this may lead to early death.

The use of stimulants can be detected by the appearance of various types of pills in well-known pharmaceutical preparations. In the use of cocaine, light powder and perhaps small snuffing spoons will be found. The individual manifests signs of tremors, anxiety, rapid speech which is rarely interrupted, sleeplessness, tremors of the hands, and perhaps chronic weight loss. Marked change in mood, either from an individual who has been rather slow and depressed to one who is exhilarated, or from an individual who has been highly exhilarated and suddenly shows signs of extreme fatigue and depression, can be indicative of the use of the stimulants.

Much publicity surrounds the hallucinogens, or psychedelics, such as LSD 25. The drug itself, lysergic acid dehydrogenase, acts primarily on the central nervous system to produce hallucinations. These are usually of a visual nature, but can become auditory or tactile under certain situations. Objects become distorted, colors seem more vivid, and sounds seem more intense. Under the influence of the drug a person can become more relaxed and peaceful, in the belief that these pleasurable sensations are an experience of "inner truths" or of "becoming more at one with oneself and the world."

The sense of acute awareness and the feelings of deep insight are distortions of reality and tend to be a result of an individual's perception of the world around him. A person's own distorted perceptions—and we all distort our own perceptions of reality—become significantly intensified. Religious and mystical feelings are commonly experienced with the drug, and quasi-religious social groups have been formed around the use of the psychedelics.

In a contrary mode, a "bad trip" is a rather horrifying experience. The "tripper" often experiences hallucinations of body fragmentation, as if parts of the inner or outer self are being separated from the rest of the body. There is an overriding fear in these instances that the "trip" will not come to an end and that the person will remain in this state for the rest of his life. The best results in treating a bad trip are achieved with the use of phenothyalizine tranquilizing agents, as in the case of other organic or functional psychosis. Many people who are familiar with the use of LSD can "talk somebody down" from a trip and achieve similar effects strictly by giving comfort and support to the individual. In actuality, the user experiences an organically produced psychosis rather than any mystical or religious experience. In recent years it has been suspected that the chronic use of LSD interferes with and causes mutations of chromosomal patterns which would later show up in the offspring of the LSD user. More recently, clinical tests have shown that there is no statistical evidence to confirm this hypothesis.

Other drugs that fall into the category of the hallucinogens include psilocybin, DMT, mescaline, a derivative from the peyote plant, STP, and many lesser used and lesser known compounds. These drugs all have similar actions, in greater or lesser intensity, to LSD. Presently there is speculation that the use of the hallucinogenic drugs may be beneficial in the treatment of autistic children by causing a flood of intense perceptual stimulations in the child. Another medical use of the drug, which at present is undergoing experimentation, is in the treatment of cancer patients who have to deal constantly with the distress of an immiment death.

Aside from the danger of a bad trip, the main bad effects of the hallucinogens are a prolonged psychotic state, severe depression leading at times to suicide, and a feeling of omnipotence which can cause the individual to perform acts that can be dangerous to himself or to others. Numerous individuals who have a history of prolonged use of LSD give the appearance of chronic brain damage when interviewed. This occurs even following prolonged drug-free states. Others, who experience prolonged psychosis, require intermediate long-term hospitalization to help them cope with their distortion of reality.

The symptoms of the use of the hallucinogens can be seen quite readily in the most careful users. They have markedly dilated pupils, obvious mood swings with a paranoid type of suspiciousness, bizarre behavior, sometimes nausea and vomiting, along with an increased pulse rate or elevated blood pressure. Other signals are irrational talking or wild accounts of hallucinations.

The drug itself comes in pills and capsules in varying sizes, colors, and shapes, which give rise to the numerous names that the drugs go under. They can be known as "white lightning," "sunshine," and many local colloquialisms to define both the color and the potency of the drug. The concurrent use of sugar cubes, which was most common when the drugs were first introduced, is no longer in vogue.

As the hallucinogens are not physically addicting and have a low potency for psychological addiction, there are usually little or no withdrawal symptoms associated with their use.

We now come to the common denominator of the drug culture—the low-level hallucinogen *marijuana*. Technically, hashish and other plant compounds are also marijuana-type hallucinogens, but will be dealt with separately. The active substance in these drugs is tetrahydrocannabinol, commonly known as THC. THC has been synthesized in the laboratory and may be sold on the street in liquid form; however, because of the difficulty in synthesizing the substance, very often what is sold as THC is actually a mixture of a psychedelic drug and a belladonna alkaloid.

Marijuana, also known as grass, reefer, Mary Jane, and other colloquialisms, is the mixture of the leaves, stems and flowers from the plant cannabis sativa. Better quality marijuana contains a higher percentage of the leaf. The potency varies considerably with the area in which the cannabis plant is grown and the percentage of the leaf being used. Hashish (hash) is a more finely extracted mixture of resins exuded by plants with different filler components, depending upon the country of origin. Most hashish is supplied from the Middle East. As it has six to ten times more THC than an equal volume of marijuana, it is exceedingly more potent. Aside from potency, however, the actions and reactions for the two substances are nearly the same.

The desirable effect of marijuana is an intoxication that in many ways is not too different from that of a moderate use of alcohol. Unlike alcohol, which is a central-nervous-system depressant, marijuana is a mood enhancer. Many external stimuli are excited under the use of marijuana, while depressed under the influence of alcohol. Music and color take on increased intensity similar to, but not as severe, as the effects of other psychedelic compounds. Under the influence of marijuana a person may find new tones and new sounds in music that he thought he was familiar with. Internal stimuli too can be greatly enhanced by the use of the cannabis derivatives. Shakespeare knew that wine increases the sexual appetite as it diminishes sexual ability; marijuana seems to enhance not only the appetite but also sexual ability. This is not only a result of a diminution of inhibitions but also an enhancement of sexual mood. Though many people may become more talkative and spontaneous, many others experience a mood of introspection and become quiet and withdrawn.

The pleasurable aspects of the drug can also be its greatest danger. Because it tends to diminish anxiety, to relieve tension, and to cause a state of well-being without any severe side effects or subsequent hangovers in the morning, many people become psychologically addicted to the drug. It is used by many youngsters as a sort of tranquilizer to help them cope with severely stressful situations. Many cannot leave the house in the morning without smoking a joint, and they may smoke three or four joints before going to bed that evening. Unlike with alcohol, very often the individual is unaware of the perceptual distortions that are taking place with the use of marijuana. He often believes that his abilities are enhanced, and he will therefore not limit his driving or other demanding activities as he would if he knew he was intoxicated with alcohol. This overestimation of capabilities in itself presents a serious life-endangering situation that most of the users do not wish to accept.

Because of the psychoactive nature of the drug, it can, in certain instances, cause psychotic-like symptoms, which can be likened to those of the more potent psychedelics. Paranoid delusions of persecution are not uncommon with even a moderate amount of marijuana. Illusions, distortions of what one sees and hears, and hallucinations, both visual and auditory, have often been experienced. In rare occasions the psychosis has been so intense that hospitalization has been necessary. Many individuals who have a basic psychotic personality *without the use of drugs* but who manage to maintain themselves in daily life without any evidence of psychotic symptoms may demonstrate full blown psychosis with the use of marijuana.

Because of the anxiety-relieving effects of marijuana, the effects of a psychotic process opposite those described above are often witnessed. There have been at least three cases in my own experience where marijuana served to alleviate anxiety in psychotic individuals, and more numerous times in neurotic individuals, allowing them to function when all other tranquilizers had failed. It seems that when they were using marijuana they were able to attend school, get to work, and form social relationships that were more positive than when they were marijuana-free. What makes this more unusual is that these same individuals are often unresponsive to the tranquilizers that presently form the armament of psychiatrists.

The symptoms of marijuana smoking are signs of intoxication, with slurred speech, staggering gait, poor perception and poor speech pattern. To a lesser degree, bizarre behavior, paranoia, or hallucinations can accompany the use of the drug. The sclera of the eyes turn red and the pulse rate is increased. Frequently, there is also an increase in appetite.

Marijuana is not physiologically addictive. Its main danger lies in its psychological habituation, in which withdrawal may include depression, lethargy, and a decreased ability to function. The immediate dangers of use were stated previously. At this time there is no concrete evidence to demonstrate that the use of marijuana *in itself* can lead to the increased use of hard drugs.

Evidence of marijuana use is obvious: rolling papers, hash pipes, an oregano-type substance with a sweetish smell, or the presence of a sweet smoke in a room. Incense is often present to cover the characteristic smoke and odor.

The drugs discussed here are the most commonly used by adolescents in the drug culture. The following drugs are those that are used to a lesser extent by the drug society and sometimes, to a greater extent, by adults who have them prescribed by the family doctor.

The inhalants include glue, lighter fluid, paint thinner, toluine, xylene and benzene. They are inhaled by "huffing," that is, by taking deep breaths of the substance, which is placed either in a paper or a polyethylene bag. The immediate effect is a sensation of general inebriation with central-nervous-system depression. Dizziness, a floating sensation, exhilaration, and a sense of well-being are all typically reported. Occasionally visual hallucinations may occur. These drugs are not used to a great extent at present. They are dangerous because they are easily used by very young drug abusers, often as young as six years of age.

The symptoms which lead one to suspect the use of these drugs are a characteristic strong odor of the substance on the breath, along with intoxication, euphoria, staggering gait, and slurred speech. The finding of numerous tubes of glue or solvent cans are obviously indicative of the use of inhalants. The main danger of these drugs is that they can be fatal. A number of fatalities have been reported either because of the high concentration of a toxic substance or from suffocation when a polyethylene bag is used to ingest the drug. Liver and kidney damage may occur with prolonged use of inhalants, along with damage to the circulatory system, to the brain, and to bone marrow. Though the drugs are not physiologically addicting, a moderate psychological habituation can occur.

Minor tranquilizers, such as Miltown, Equanil, Librium and Valium, are actually more often abused by the adult world than by the adolescent population. They are, in many ways, similar to the barbiturate hypnotics described above, but tend to have a greater effect upon the relaxation of the skeletal muscles and on the relief of anxiety with less drowsiness. They are used medically as antianxiety agents, mild sedatives, or muscle relaxants.

Physical dependence and addiction does occur, and it is often necessary to detoxify an individual in a manner similar to the way one would with the barbiturates. Though convulsions have occurred through sudden withdrawal, death or serious injury at such a time is unlikely. The main dangers, along with an addiction, are the effects of the intoxication and the fact that overdose may lead to death. As these drugs are often prescribed to psychiatric patients, they can often lead to suicide.

The major tranquilizers, Thorazine, Stelazine, Mellaril Haldol, and Sinequan, are used psychiatrically for the treatment of psychosis. They produce an effect not too different from that of the minor tranquilizers and the barbiturate hypnotic group.

They tend to be less dangerous, however, as they do not cause a physiological addiction, and death by overdose is exceedingly rare.

The main dangers include convulsions with high dosage, secondary effects of intoxication, and liver damage with prolonged use. The symptoms that one sees are extreme drowsiness, slurred speech, muscle stiffness with protruding tongue or the retraction of the eyeball upward, and generalized symptoms of intoxication.

Only a doctor, of course, can adequately diagnose a drug condition or relate symptoms to a specific drug. Nevertheless, parents cannot ignore the medical danger signs described here, or try to rationalize away their severity. Nor can parents discuss drugs intelligently with their children without a basic knowledge of the facts outlined here.

—10—
The Drive-Defense Complex in Drug Use

The primary drives that are associated with drug use or abuse are on the oral level. In order to appreciate the importance of this fact, we should examine first what we mean by "oral."

The oral level of development is the very earliest stage in human life. It is the stage at which satisfactions come primarily through oral activity. The mouth is the all important organ of gratification, and is referred to by Freud as the primary erogenous zone. Prior to the second year, the need to be fed and the satisfaction of the hunger impulse are basic to every other activity. The infant knows little more than the desire to have that hunger satisfied as soon as possible. There is little or no frustration tolerance; there is no self-assertiveness beyond the seeking of the gratification of the need. At the same time, the child avoids any alienation from the person who is fulfilling this need. The child is unconsciously aware of his total helplessness and dependency on another human being's whim, and soon learns that any action that may alienate that person becomes dangerous. The fear of alienating the person on whom the child is dependent is markedly increased whenever that person is negligent in the fulfillment of this duty of child rearing. The child perceives that he is on dangerous ground if he were to displease the parent or parent figure.

It is clear that the frustration of oral needs has a direct bearing on the security of the child. The anger towards the parent, or primary dependent object, for not nurturing his needs and the *fear* of the anger because it may turn that person away both feed the insecurity and conflict in the infant's life. He begins to fear that, rather than a few of the needs being fulfilled, there will be none at all. When oral needs have been incompletely fulfilled during the first two years, they grow in intensity in childhood and throughout the person's life.

No one, no matter how free from neurotic conflict, is ever completely devoid of oral needs. In those whose fear and insecurity were formed in infancy, however, the oral need remains the paramount objective for gratification throughout life. It is true that, as the child matures, there is a natural tendency for diminution in the gratifying of oral needs by others. Parents come to expect maturation and subsequent independence in the child. By the time the child reaches adolescence, society also exerts an expectation for a certain degree of self-reliance and independence. Along with this the child has certain expectations and awarenesses of himself that place further demands upon his self-reliance. Nonetheless, the oral need, if never fully gratified, is still predominant in these adolescents.

Because of the constantly increasing frustration of oral needs there is a subsequent increasing rage—and accompanying depression. The mechanisms which now become predominant are not difficult to understand. Underlying the depression is always the anger caused by oral frustration. The fear that the parent will be lost leads the child to suppress his rage. In turn, the child projects rage onto his parents and becomes fearful of reprisals by the parent in the form of abandonment or isolation. A vicious cycle results in which the fear of reprisal increases the degree of anxiety, which must eventually be repressed.

The imagined reprisal expected of parents may take many forms aside from alienation. Frequently the child's own violent wishes toward the parents are projected onto the parent and are felt to be an appropriate punishment that the child feels he deserves. This feeling is further increased by the primitiveness of the child's superego, as this area of the psyche fails to develop to maturity.

The child projects the parental image in toto into his superego, his emphasis on orality causes the incorporation of the parent there, and the fear of the parent is experienced as guilt. The incorporation leads to a secondary aggressiveness because the child views the negligent parent as an aggressive individual and the negligence as a hostile act. In short, the primary oral drive of the adolescent is frustrated by the subsequent rage which alienates the parents, by the fear of the results of the

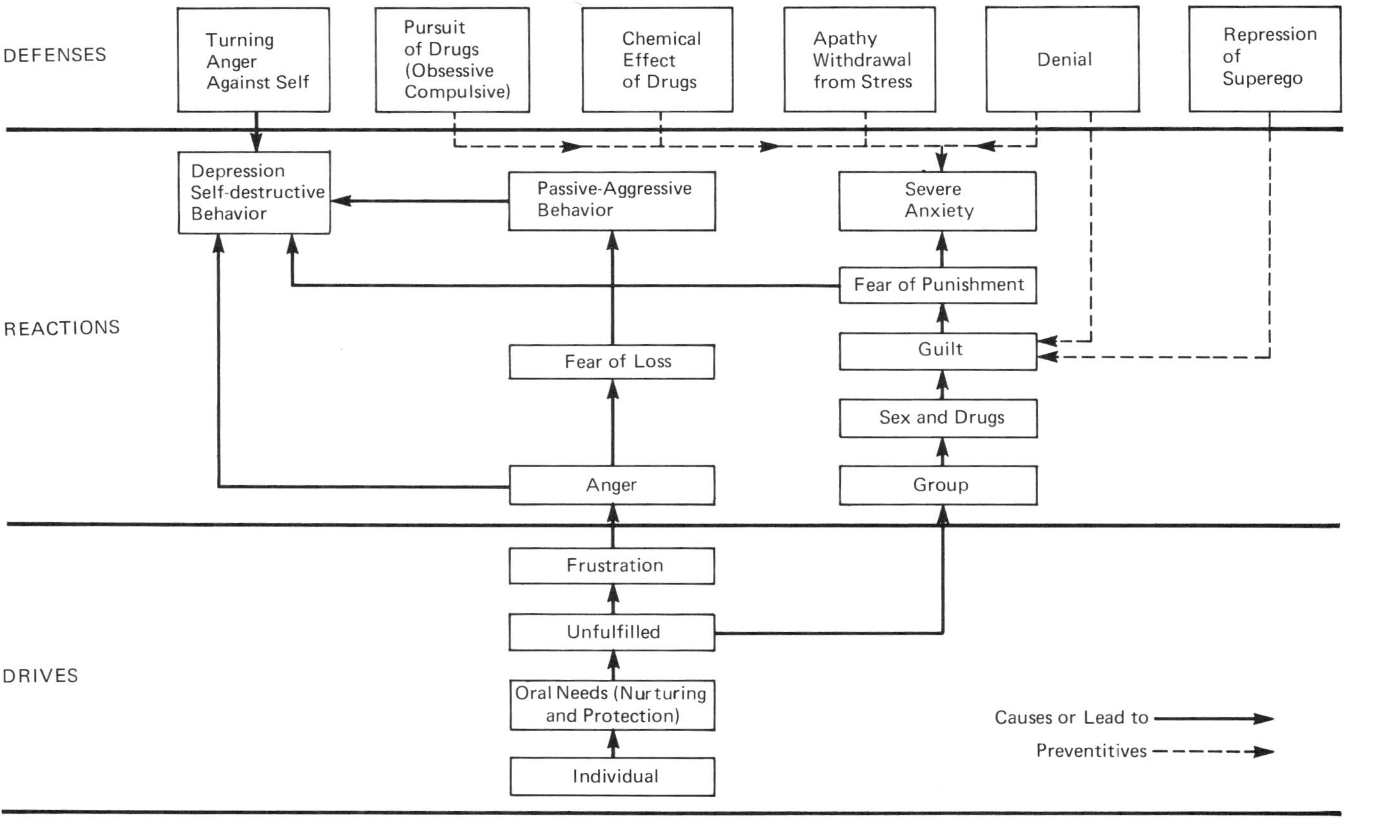

Drive Defense Complex

rage, and by the constant anxiety over the anticipation of alienation *from* parents or punishment by them.

As the child matures and as his own concept of his need for independence grows, the child's feelings of inadequacy grow too. The basic self-image is of an infantile individual incapable of coping with the adult world because of arrested development in the oral stage. However, as a result of his rageful feelings, a series of defense mechanisms take over, which prohibit the child from dealing with the world. There is a confusion in the child's mind between being assertive and being aggressive; the need to repress any aggression subsequently represses the assertiveness. As we have seen earlier, the child with an inadequately developed ego will maintain the feelings of omnipotence that are present in early childhood. The child, therefore, becomes frightened that his rage will magically destroy the very people he most needs, namely, his parents. He has developed ambivalence towards the parents, the simultaneous feelings of hate and love. The love is, of course, based upon the need to be fed and the hate is based upon the feelings of not being fed. The negative aspect of this ambivalence and the feelings of omnipotence lead to the repression of all aggressive feelings.

As the aggressive tendencies are repressed, the child is literally afraid to assert himself at all. He sees assertion as dangerous as aggression and therefore remains in a passive state. He develops an underlying passive-aggressive personality, that is, where all of his "aggressivity" is taken out in a passive manner. His behavior is like that of a child who refuses to open his mouth when the mother tries to feed him, or the older child who refuses to go on a picnic to punish the rest of the family. Only passivity is seen by him as a safe expression of aggressivity, because it is not felt to be destructive. The passive-aggressive nature is one of the underlying character structures of the drug abuser and plays a dominant role in the self-destructive need of the individual.

As the child continues to develop emotionally—while still being arrested, for the most part, in the oral state—and wrestles with his feelings of independence, he will begin to become aware of feelings in the genital area. In the normal child this awareness corresponds to the development of a normal heterosexual relationship. Since the orally fixated child sees everything in the terms of being fed, confusion begins to arise in the psyche between his desire for closeness with the maternal figure and his genital sexual feelings. Both male and female children are affected: the male becomes afraid of the incest taboo, while the female becomes confused about homosexual feelings towards her mother. They, therefore, cannot allow themselves to remain close to their parents and they start breaking away in a pseudo-independent manner. Because of the pressure to make the break at such an early stage, they are slow to develop an adequately mature identification with the parent of the same sex, to allow for resolution of their oedipal or homosexual feelings. There is, therefore, a constant lack of identity which furthers the need for a *dependent* relationship. On the other hand, they cannot follow in their father's or mother's footsteps because of the anxiety over their sexual conflicts.

In the adolescent this anxiety is manifested in a number of ways. The confusion between dependency and sexual attraction remains and becomes a stimulus for sexual "acting out." In speaking to youngsters who have acted out sexually, one finds that, to a great extent, very few ever really enjoy mature genital sex, and yet, most are seeking the closeness of another human being. They attempt, through intercourse, to seek a fusion with the maternal figure, which can be either a male or a female. For their own sexuality has never matured to a point of differentiation, and everyone is seen as a mother. This sexual confusion accounts for the high level of homosexual feelings among many drug abusers.

In adolescence there is a resurgence of intense sexual feelings. As they come to the fore, the natural craving for parents is confused with a sexual desire for the parents. It is not unusual to hear overt fantasies of incest with the parent of the opposite sex or even of desires for homosexual activity with a parent. These sexual fantasies are, once again, not genital but oral, since the youngsters lack the need for fusion that they desired as children. However, the incest taboo has produced so much anxiety over these feelings that there is an intense pushing away of the parent at this stage in life. As the need to push the parent away becomes intensified, there is once again an alienation from the primary dependency object and an increase of anxiety. The adolescent then turns to his peers for the gratification that he seeks and displaces the oral sexual needs from the parent to a member of the peer group.

Because of the displacement of so many of the drives from

the normal channels into symptomatic channels of behavior, the defensive nature of drugs may be considered to be based on a symptom complex. Underlying the use of the drugs are inevitably depression, anxiety, guilt, lack of identity, feelings of worthlessness and of inadequacy. These weak, repressive mechanisms are bolstered by the entire armory of defense which drugs promise.

The intertwining and complexity of the defense mechanisms of drug abuse are what complicate the treatment of drug abusers. At each stage in the development of drug abuse, a repressive mechanism plays its part. There are basically five defensive functions of drugs only one of which is directly dependent upon the chemical component. First, there is the immediate chemical effect that the drug produces, which is, as often as not, self-tranquilization. Second, there is group interaction and what it promises for the individual. Third, the self-destructive force of the drug, a constant psychological partner in depressed individuals, becomes increasingly important in the drug-abusing personality structure. Fourth, the pursuit of the drug itself takes on many diverse meanings to the drug abuser. Fifth, there is a rationalization mechanism developed in drug abuse which promises relief in other areas of living.

THE GROUP AS A DEFENSE

It has been mentioned many times previously: the single most important factor leading an individual to a life of drug abuse is an unstable parent-child relationship. Despite his mistrust of parental figures, however, the drug abuser is not at all ready to abandon his oral strivings. He therefore displaces his need for dependency from the mistrusted parent onto a peer group.

The group, rather than an individual, becomes the central focus of his libidinal attachment. He essentially avoids the risks that one runs in an intense love relationship with one individual. Having learned that he cannot trust an individual to fulfill his needs, he scatters his desires throughout the group. Furthermore, by putting his dependency needs on a peer group, he makes a compromise in his dependent-independent conflict. He can justify his dependency on his peers by denying that he in any way depends on the *adult* world; at the same time he gratifies his dependency need. What is it that makes a *drug-*

abusing group so desirable to such an individual? Why is it that he cannot seek a more productive, more healthy peer group? The group has often been blamed both by the individual and his parents as the reason for the child's taking drugs. However, it is the child who seeks this group and it is not the group that seeks the child. The reason for the child's choice lies in his needs—needs which would make him unacceptable in a more productive group of friends.

Primarily, the drug-abusing group is made up of individuals who place absolutely no demands on a new member of the group other than the joint activity of taking drugs. The child has no self-esteem, no self-worth, and an overwhelming feeling of inadequacy; the lack of demands by the group accordingly becomes of paramount importance. Because the standards of the group are so low and so easy to meet, there is little anxiety produced in the company of these friends. There is no need to act in a productive manner, as in a school setting or in a job, so the feeling of inadequacy wanes. The only demand frequently made is in the area of sex for young girls, yet even this is not consistent.

In the psychic structure of the drug abuser's character, the severely punitive and infantile superego is always present. This superego would produce intense feelings of guilt and fear in the individual's dealings with his peers *outside of the drug group*. Because of the antisocial nature of this group, however, the individual is able to find a new image upon which to build his ego-ideal. Among the most popular forms of this ideal is the antisocial youngster who rejects parental values. It becomes acceptable, at least superficially, to steal, to disregard all authority figures, to remain truant from school, and to engage in sexual activities. These activites fit well into the need of the usual drug abuser to identify with a drug group: he does not feel the guilt nor the fear of punishment that would be felt in any other situation.

Another common ego-ideal that is formed from the relationship with the group is that of the hard-core addict or jailbird. The image of being "manly" or "cool" in the face of risk is highly desirable in the drug addict crowd. The addict will often boast about the amount of heroin he is able to inject and will often lie about the quantity of bags that he is addicted to. The hierarchy of the addict world is defined by the number of bags of heroin that can be mainlined during the day. Similarly, an abuser will boast about the number of years that he has spent in jail and the degree of severity of the crime. It is not infrequent to find a youngster who has been placed in jail on a possession misdemeanor boasting to his fellow inmates that he had been sent up on a felony rap for car theft or murder. Such boasting obviously stems from the need to overcome feelings of inadequacy involving one's masculinity.

Heroin diminishes the sexual ability of most men. Part of its desirability is therefore its reduction of sexual strivings among men who are plagued by feelings of masculine inadequacy. Indeed, most addicts can be expected to be suffering from a weak masculine identity. Many come from homes in which the father is either absent or is an exceedingly weak figure, where most of the child rearing has been left to the mother. The heroin provides a rationalizaton for sexual impotence. The "jailbird image" then becomes the source for a new masculine identity. Many an addict would rather give up heroin, paradoxically, than this image. It is often a major part of the treatment of the addict to break through the jailbird image and to allow him to eventually admit to himself the acceptability of attempting a straight male image.

The third most common ideal that the young drug abuser seeks is that of the totally apathetic individual. He convinces himself that nothing matters and that nothing is important. He identifies with "drifters" who have generally removed themselves from their homes, to live on a meager subsistence, arrived at through begging, and during the warmer months, as often as not, live in the streets. Unlike alcoholics who live isolated in the Bowery, these youngsters place heavy reliance on group dependencies. Being so young and still so close to parental influences, they find it hard to move out on their own in a totally independent manner. Furthermore, they need somebody to reinforce their rebellious attitude against parental values.

The apathetic stance which this type of abuser has developed is necessary to protect him from the severe anxiety that comes with constantly desiring and striving for something which he is afraid that he can never attain. This is especially true when that which he feels must be attained is basic for his survival: food and shelter. It therefore becomes necessary for him to set up a new society, separate from striving and separate from productivity in order to fulfill these needs. Furthermore, the youngster is wrestling with a severe "reaction-formation" to his own intense hatred of the absent dependency object—his parents. The apathetic mode of existence makes any aggressivity unnecessary and therefore he may keep his aggressive impulses more deeply buried and more stringently under control. In other words, in order to avoid saying "I resent my father," he says "Well, my father doesn't really matter."

Another type of group to which the drug abuser *may* gravitate is one very active in social rebellion: anti-war demonstrations, anti-government picketing, and struggles against racial injustices. Drug abusers who choose such outlets are functioning on a much higher ego level than those in the previous groups, and so are far less threatened by their own aggressive impulses. On the surface they appear to be healthy, normal youngsters pursuing causes that a large segment of the adult world deems justifiable. They arouse the sympathies of the less active population, and often bring about positive social improvement that otherwise would not be achieved. They have been involved in civil rights' marches in Mississippi, anti-war demonstrations at the Pentagon, and were influential in the McCarthy phenomenon of 1968.

Though they too cut school, behave in antisocial ways, and show a marked falling off of socially acceptable productivity, their cause is so just or their self-righteousness so assured that it can mask an underlying drug-abuse problem no different from any other. It is not the end result of their behavior that is of concern, but how it may disguise a different motivation from one of altruism and good will. Let it be doubly underscored that we do not mean to imply that all of the youngsters who are involved in these activities are there out of unhealthy motivations. This group is usually limited to those of a high intellectual level and a corresponding self-image that they would like to maintain. They come from a cultured environment, with parents who are active and concerned with the community, very often at the expense of their children. It is frequently the *community* activity that pulls the parents from the home and is the source of anger of their children.

The anger that is felt towards the neglectful parent is displaced by the child from the parents to society. As these parents are rarely "bad" parents, it is difficult for their child to justify his anger towards them. The parents themselves generally show concern for the child in an indirect way, but more frequently they are concerned for the child in terms of their own needs. The child therefore takes the anger that he feels towards the parent and displaces it onto social injustices that he sees around him. He aligns himself with a specialized group of individuals who are politically aware and active, and secondarily *who are rebelling against all of the standards his parents have identified with*.

The most commonly used drug in this group is marijuana. Rarely do these children go on to harder drugs, and, if they do, it is most common that they use them sporadically rather than on a regular basis. Marijuana is experienced by this group as a symbol as much as a defense: a rebellion against the social ills that the government stands for. They intellectually justify its use by this rationalization, and coincidentally justify violent behavior on the same grounds.

When one recalls the 1968 Democratic Convention in Chicago or the anti-war protests at the Pentagon, it is clear that the most frequently expressed emotion was anger, often leading to violence *on the part of the protestors*. The primary object of this anger was not always the provocation of the cops in Chicago or the National Guardsmen in front of the Pentagon, but rather a parental object who had been absent from these children for a good part of their formative years.

Let us go into detail in the case of N who was discussed briefly previously. N, seventeen years old when he entered treatment, was a junior in high school and up until that year was considered highly successful. He was captain of the basketball team, president of his class, and his marks were consistently above 85. In the middle of his junior year he began to demonstrate attitudes that were displeasing to his parents, and eventually he became more and more depressed. He joined the Black Panthers—the only white member of the organization in that area. He became a leader of antiwar movements in New York, and an important co-ordinator for the McCarthy candidacy. During this time he started to become heavily involved with marijuana.

His marks fell precipitiously and his whole value system changed. Previously his goal had been to go on to college and enter the field of law, but at that moment he was totally undecided as to which course to pursue. He felt that law would be hypocritical, as it meant moving into the mainstream of society. He wanted to have nothing to do with the social order that bred prejudice and war and violence. During that year he went to Washington to participate in the Pentagon riots, and his head was smashed by National Guardsmen in front of the Treasury Building. He became involved in various fights because of the Black Panther connection, and constantly justified this "rage" by claiming that he felt it promoted the social good. Fortunately, he still maintained many of his old values, but as he found himself slipping from these values he developed more conflicts and became more and more depressed. It was the depression that led him into therapy.

As therapy progressed he became aware of feelings that had been with him since his early childhood. He recalled feelings that his parents were never really interested in him as a person, but only for what he could do for them. He recalls hearing his mother say that he was not a planned child; from that time on he felt as if he were a burden, and the only way he could be accepted was to be highly successful. From his earliest years he had been able to maintain his success in terms of his scholastic achievements and numerous athletic awards.

He had two younger brothers, both of whom, he felt, were less gifted than himself. Superficially he believed that his parents loved him. It was only after months of uncovering in therapy that he became aware that he felt his parents loved only his achievements. It now became clear that he had harbored intense angry feelings for both his parents and for his younger brothers, who were intrusions upon his life. He recalled flailing out in blind rages at his middle brother; he recalled incidents at school where he would hit other children without any provocation, and some earlier sexual experiences which contained marked elements of sadism.

Further therapy helped him understand the feelings that were brought about by athletics. On the basketball court he was able to alleviate his aggressive impulses in an acceptable way, for on the surface he had always been nonviolent. He expressed the loathing of violence in any form, and became particularly angry with himself when he unleashed his moments of fury. He was a pacifist when it came to the Vietnam War, and said he would have refused to serve because he refused to kill. All of this contradicted his unconscious fantasy—which contained much violence and killing.

These sadistic fantasies were loathsome to him. He therefore had employed defenses early in his life: reaction formation against violence and displacement of the violent feelings toward his parents onto athletics. When these defenses were not strong enough to keep his rage intact, however, his violence would break through in an unprovoked attack on his brother or schoolmates.

In his seventeenth year he became acutely aware of having to make a break from his home. He had always appeared to be a highly independent individual, but he began to fear the break from his parents. He was unsure how his parents would feel towards him if he left home; he began to manifest anxieties as to whether or not they would accept him back when he returned home from college. The previously employed defense mechanisms were no longer strong enough to maintain his rage on an unconscious level, nor were they strong enough to maintain his dependency needs on an unconscious level and let him lead a life of pseudo-independence. In order to reinforce these defense mechanisms, therefore, he had to employ massive denial of parental relationships and he had to seek a more direct expression of his anger. He made a total break from his parents' values and standards, displacing his anger onto the social ills which he could accept as a justifiable target.

With therapy, over a period of months, these feelings became clear to him. He began to see his anger towards his mother for being what he considered neglectful of his needs. He was able to understand and accept his anger towards his middle brother who, he felt, had replaced him in family affections as an infant. As he recognized these feelings, his rebelliousness became less and less necessary; by the middle of his senior year he had redirected his energies back towards a college career. At this time he had developed a strong identificaton with his therapist and his goal at the time was medicine.

He was able to separate himself from the group to which he had belonged. He was no longer in need of people to reinforce the justification of his vengeful feelings. He turned from political activism and returned to some degree to sports. Happily, he still remained active in politics, but in a less violent way. He disassociated himself from most groups, and turned more to himself and to his family. At the end of his senior year he was able to go off to college with plans for premedical studies.

Following his first year in college he realized he was able to adjust academically, but other problems arose. He still maintained basic sado-masochistic relationships with his girlfriends. He found that he was sexually impotent or at least deficient. He became frightened by these feelings and decided to take a leave of absence from college for a year to return to therapy. During the following fourteen months in therapy the rage towards his mother became more apparent. Much of the anger towards her was expressed directly because of his inability to channel it. He became acutely aware that the anger towards his mother now was being displaced onto other women, either directly through sadistic actions or indirectly through a reaction to having intercourse in which he refused to satisfy them.

He also viewed sexual intercourse as an act of aggression. He confused in his unconscious any act of aggression with violence towards his mother. Coincidentally, he had also projected his own rage onto the women with whom he was having sexual encounters, and so became fearful of *their* anger towards him, with active fantasies of fear of castration. Once the rage had diminished and it was no longer necessary to displace it from his mother to other women, he began to have satisfactory intercourse.

It was at this time that he was able to give up his marijuana use entirely. The drug abuse had gone from a social revolutionary act to a tranquilizing act to help him participate in sexual relationships. He returned to school following that year of therapy; he was able to maintain a relationship with a young woman both socially and sexually without the sado-masochistic component that was previously present in all of his relationships. He was also able to resolve the identification with the therapist and decided to go back to the study of law.

Though most of this case history places the emphasis on the defense of the individual, it is necessary to restress the role that the group played in the early part of his life. He sought out a group of intellectual, politically active individuals, who themselves had needs for violent action to gratify their own internal desires. Through this group he was able to justify his own violent needs and redirect them from his home outward to society. He would have been unable to do this on his own, as he needed the reinforcement from a family surrogate in order to be able to justify his previously loathsome violent actions. Because of the socially rebellious nature of this group, it was using marijuana extensively as a part of social rebellion.

It became evident later that N's use of marijuana also played other roles in his psyche. It served as a tranquilizer to subdue some of his passionate anger, as a sexual stimulant, and simultaneously as a means to rationalize his sexual inadequacy, which he would excuse because he was "stoned." It is clear, however, that marijuana did not have to be a part of this entire psychic picture; his involvement with an antisocial group would have just as easily served his psychic needs.

THE DRUG AS A DEFENSE

The most commonly desired effect produced by drugs is from the barbiturates and the amphetamines. There is an antidepressant effect from these drugs, the latter of which has been used for years by the psychiatric profession in order to alleviate acute depressions. This antidepressive effect is brought about by a direct stimulation of areas of the brain causing feelings of an unrealistic euphoria which masks the depression. Barbiturates, on the other hand, are basically a depressant. However, they appear to have two chemical effects which are not too different from those experienced with alcohol. If not taken in too great amounts, barbiturates seem to chemically suppress rage which, if the level of barbiturates is increased, then breaks through in intense violent behavior. However, most of the bariturate users limit their intake to a level where they are too numb for violent, emotional upheaval. As the rage is suppressed so is the need to turn the rage against oneself, and therefore, the depression is diminished by cutting off the id drive which is causing it.

Secondly, as with alcohol, barbiturates diminish inhibitions and feelings of guilt, both of which enhance the depression. The intake of barbiturates by the individual is not too different from the prescription of barbiturates for agitated, depressed people. It has been observed lately that the barbiturates and other depressants, like methaqualone, have become the primary drug of choice for most drug abusers, and this has been confirmed by governmental studies. The depressants are so popular because

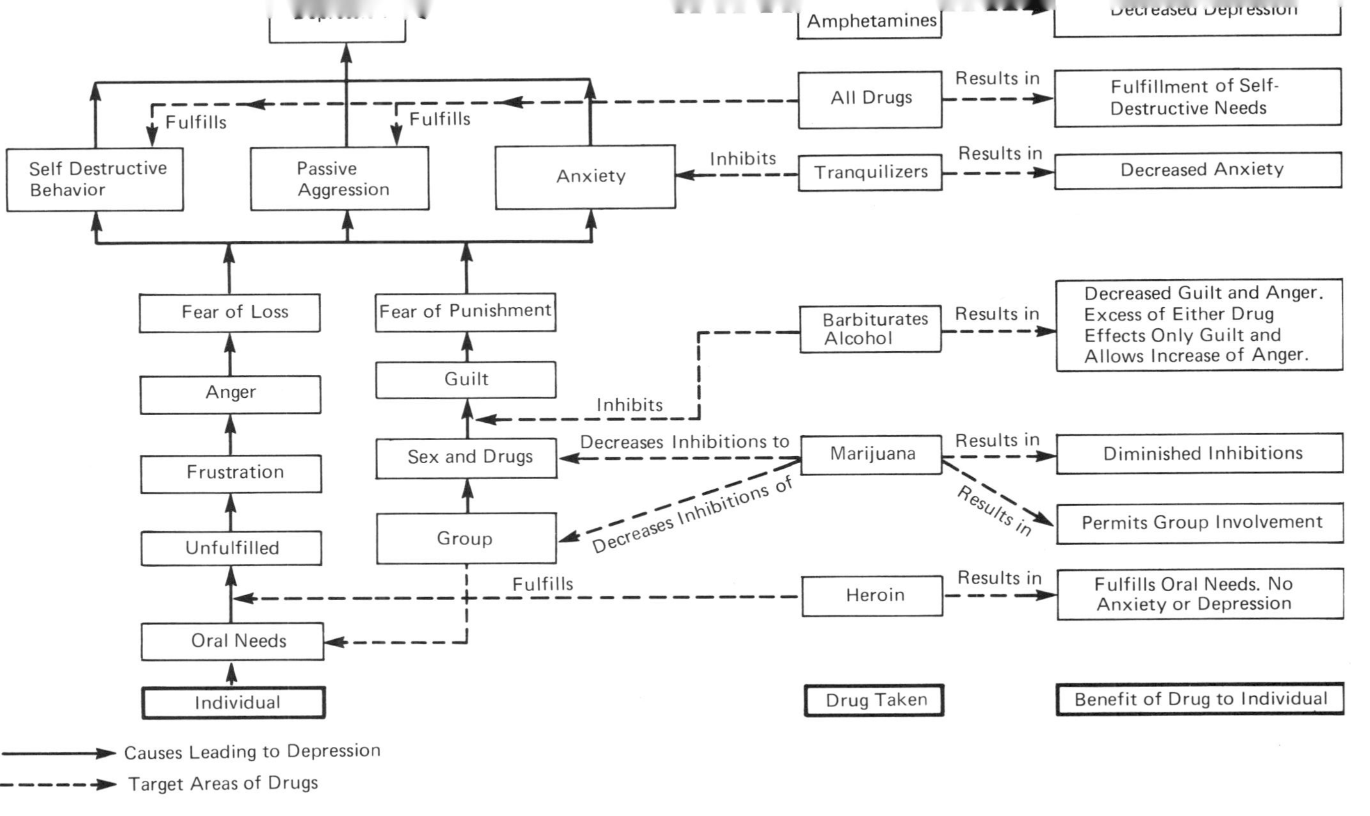

Psychological Benefits of Some Common Drugs

of the number of people with underlying depression which the drugs diminish. Aside from combatting depression, many of the drugs have a strong antianxiety effect similar to tranquilizers. The drugs on the street that are most effective against anxiety are the depressants, alcohol and heroin. For years phenobarbitol had been prescribed as an antianxiety agent prior to the advent of the tranquilizing drugs.

As had been stated previously, the primary drive causing anxiety is the strong aggressive drive secondary to the anger that has been brought about by the frustration of the oral need. The chemical effect of heroin, alcohol and barbiturates causes a mental state which diminishes oral drives and leaves the individual with a numbed effect requiring nothing more than drifting off into sleep. There is a diminution not only of the aggressive drive but also of the sexual drive, which frequently causes its own anxiety. With the numbing of all the senses, emotional pain, along with the physical pain that these drugs were originally intended for, diminishes. There are no desires, no caring and no striving, bringing about total apathy. Because of the apathetic attitude, the anxiety produced by the fear of not being able to cope with the activities of life diminishes. There is no more efficient way of diminishing anxiety than by the feeling that nothing matters.

With heroin there is also a primary gratification of the oral need. It has been stated that the heroin injection is symbolic of a penile intrusion with the subsequent rush which has been likened to an orgasm. This cannot be entirely true because of the basic oral nature of the heroin addict. What is more likely is that the penetration of the needle into the vein is more symbolic of the penetration of the mother's breast into the mouth, and that the rush that is caused by the heroin is less like an orgasm of a genital nature and more like the rush of warmth when the baby swallows the first drop of milk and the hunger pangs are relieved. When the heroin addict is deprived of his heroin, his tantrumlike behavior is far more similar to that of an infant deprived of food than of an adult deprived of sex. The whole basic character disorder underlying heroin addiction is one of such an intense, passive, dependent nature that the individual rarely has progressed beyond the very early stages of development. There is nothing that they want more than to be taken care of, comforted and fed, both by their drugs and by the people whom they choose as companions. After the first rush of heroin has been enjoyed the addict goes "on the nod" and describes a type of sleep that is similar to what one observes when an infant falls asleep after a meal. The entire procedure of taking heroin from the procurement of the drug to the final "nodding out" is similar to the stages of an infant feeding. Therefore, heroin not only diminishes the aggressive drive through direct chemical anti-aggressive effect on the brain, but also diminishes the need for the aggressive drive through the gratification of the oral wishes.

Many young drug abusers are suffering from psychotic disorders. There seems to be few legitimate tranquilizers that are as antipsychotic as heroin in some cases. Heroin addicts have used it to mask the psychotic symptoms of schizophrenia, paranoia and depressive psychosis. It seems to work not only by diminishing the anxiety and therefore diminishing the psychotic symptoms, but also by chemically alleviating the psychosis itself. In the case of one young man, his paranoid schizophrenia was totally masked when he was using heroin. However, when the heroin was discontinued and he was left to face stressful situations without its benefits, all the symptoms of his paranoia became evident. He had delusions of persecution, had occasional hallucinations, and had severe rage reactions that often led to violence.

Other drugs that may have antipsychotic effects are the barbiturates and marijuana. The barbiturates in psychosis are effective more in the alleviation of the psychotic anxiety than they are in the reduction of symptomatology. However, much of this anxiety is alleviated by the assistance of the weak ego in suppressing id impulses that generally break through to the surface. In the quelling of both the sexual and the aggressive impulses the barbiturates aid the psychotic to function with decreased stress and increased efficiency. Apparently it is not only the barbiturates but the other depressants, such as methaqualone, that have these antipsychotic properties. In the case of J, he was suffering from a severe psychotic depression. While he was under the influence of barbiturates or "quaaludes," he was able to maintain a level of functioning which permitted him to be gainfully employed and to attend school. With the discontinuance of the drugs, because of the fear of violating probation, his anxiety level increased so greatly that he was no longer able to endure the stress. During therapy the phenothyazine tranquilizers, alone or in combination with other antipsychotic and antianxiety tranquilizers, were prescribed with little or no effect.

Marijuana is usually thought of as a drug that can cause psychotic symptomatology. However, in at least one case it was used by a schizophrenic boy in order to diminish his anxiety enough to allow him to function in school. While he was under the influence of marijuana he was able to attend school, be attentive to the lectures and to receive decent marks on the examinations. When he was no longer smoking marijuana, he was unable to endure the school pressures, and his marks and attendance fell. It is not to be implied that marijuana is an antipsychotic agent, but it can alleviate the anxiety to a degree that it allows functioning in an individual who is overwhelmed by anxiety. Many youngsters who have never had the benefit of psychiatric treatment have medicated themselves with the above mentioned drugs, making a desperate attempt to alleviate the psychotic anxiety which is so intense that it prohibits all functioning and activity.

Another paradoxical situation is with the use of LSD among paranoid schizophrenics. In interviews with a number of heavy users of LSD, it became apparent that they experienced hallucinations prior to the use of the drug and that these hallucinations were frightening because they were not understood. During the LSD experience the psychotic youngster experienced a certain feeling of control over the previously uncontrollable hallucinations. He was able to reason that these hallucinatory experiences were a result of the drug and therefore they became less frightening. Anyone who has worked with people who are suffering from schizophrenia is aware of their strong desire to control their own psyche. Feelings of loss of internal control are extremely frightening. The striving for the maintenance of the psychic equilibrium is carried out with the use of LSD. It in no way alleviates the psychosis nor allows for better functioning but it does alleviate the internal turmoil of the individual causing many people with psychoses who have experimented with the drug to find it desirable for their particular psychic need.

THE WILL TO LOSE

Without a doubt, the most dangerous aspect of drug abuse is the self-destructive nature of the problem. Unlike suicidal individuals, drug abusers are oblivious to their self-destructive activities, and therefore are free of pain and unwilling to seek help.

Nonetheless, there is little difference between the motivation for a suicide attempt and the motivation to maintain oneself in this destructive life style. It is because of the need for self-destruction by many drug abusers that educational programs reach only a small percentage of students, who, in all probability, would have turned from drugs on their own after a brief period of experimentation. For hardened drug abusers the fact that drugs are dangerous is part of the unconscious need.

The defense mechanism that is typically employed by drug abusers to continue the use of drugs is *denial*. It does no good to argue that anything will happen to them; they assert they know what they are doing, and that drugs are only dangerous to the other person who "happened to overdose last week." Unconsciously, however, the drug abuser knows that drugs are dangerous and *part of his motivation in seeking drugs is that very danger*.

What is there to be gained by the drug abuser with this self-deception? As we have seen in the discussion of depression, the primary motivation of drug abusers is that of turning anger against an introjected object, the parent. To put it more simply, they turn anger against themselves in order to avoid destroying that person on whom they are so dependent. The unconscious wish is to destroy the other individual, but, as this is so unacceptable, the destruction is played out against oneself.

In a group therapy session that was conducted with eight drug abusers, two out of the eight had active fantasies of destroying their parents. These fantasies were totally unacceptable to them. Another four out of the eight had thinly disguised fantasies of destroying their parents, which were brought to light in eight to ten group sessions. Only two were not aware of feelings of wanting to destroy those very people on whom they all felt so very dependent. For all of these youngsters the fantasy was overwhelmingly frightening. As much as they wished their parents dead, not one of them could move out of the house and live, even with another family, for even a short period of time. The two members of the group who had active fantasies of destroying their parents had also made suicide gestures concurrent with their drug abuse.

Another became acutely aware of many of the self-destructive actions that she would employ, constantly hopeful that the parent at whom she was angry would take notice. Once, when one of the therapists had asked her to perform a task which she found odious, she found herself wandering in the street in the pursuit of that task—hoping that she would be hit by a car so that the therapist would feel guilty for having asked her to undertake the job. This is the usual situation in many groups. Drug abuse serves a purpose similar to that of being hit by a car.

The fantasy behind many self-destructive acts is the idea that by destroying the internalized object, one is indirectly destroying the object at whom the anger is directed. The guilt then felt by the other person will secondarily gratify the punitive wish. The fantasies revealed in therapy about the results of parental discovery of drug taking are inevitably a combination of facts which contain *one constant theme. The underlying feeling is that parents will feel terribly guilty at what is happening to their child and alter their way of life to bring them closer to their child.* There is inevitably the desire to cause the parent pain and grief over the loss of their child. Very often, active death wishes are felt by children, as they fantasize the family standing around the grave, weeping bitterly and wishing that they had treated their children differently. Youngsters feel gratification at such supposed parental suffering; at the same time they express a feeling of helplessness about bringing about this suffering in any other way.

Unfortunately, many parents who have been neglectful throughout the child's life now begin to feel guilty over what is happening to their child, and therefore become easy prey for the attack of the "loser." The more parents become agitated and flustered because of their child's failure, the more the child is tempted to continue this behavior. On the other hand, if parents pay no attention whatsoever to the child's action, he will continue to make more and more serious attempts to disrupt their equilibrium and cause suffering.

One young woman in our experience consistently left messages to her parents about her behavior. They would find her birth control pills and her rolling papers day after day. Not knowing how to handle this situation, they did not react. As time went on they began to find barbiturates, and they stopped finding birth control pills. Their daughter became pregnant and asked for an abortion, for which the parents were asked to pay. Still being unsure as to what course to follow, the parents consulted a social worker. She advised them to avoid getting entangled with the child's acting out, to try to deal with her rationally. Though this was wise advice, the parents interpreted it to mean to ignore what the child was doing. As the ignoring continued, there was a rapid succession of three barbiturate overdoses, each requiring hospitalization. The final overdose was so serious that the girl remained in a coma. Still not receiving the desired response from her parents, she later turned to heroin.

During this entire time the daughter vigorously denied to her parents any desire to have anything to do with them. She hated the idea of being dependent upon them. She refused to allow herself to form any relationship with them because she was so angry about her childhood neglect. It took three years of psychotherapy to make this girl somewhat aware of what her behavior meant. It became apparent to her through her memories of her early childhood that she was not raised by her mother but by her grandmother; her mother was fearful of "making a mistake" with the child. When she was six, a brother was born—but now the mother had to take a child-rearing role because of the grandmother's age. This so angered the daughter that she harbored a resentment against her parents throughout her life. She had to deny to herself the desire for parental involvement in a pseudo-independent manner.

She constantly needed to punish her parents because of the neglect that she felt at their hands. Her self-destructive fantasies began to unfold throughout the course of therapy: the distress she caused her parents was her way of punishing them. Simultaneously, her fantasies included the desire for her parents to put a stop to her behavior. *She had maintained a remnant of the childhood fantasy of her parents' omnipotence.* Unconsciously, she believed that no matter what she did, if her parents loved her enough, they would be able to put a stop to her behavior. Once she was able to accept her desire for her parents' love, she was more ready to accept her anger towards them. Once she could accept the anger, she was better able to externalize it and no longer needed the self-destructive route in order to punish her parents.

Throughout the early part of her therapy, this girl would drift in and out of the therapeutic situation, often disappearing for months at a time because she was too frightened of her own dependency need of the therapist. Once she was able to accept her desire to be taken care of, however, she did not have to leave the therapy situation every time that she felt dependent. Concurrently, she was able to form a relationship with her parents that ended her periodic moving away from home.

This case exemplifies not only the desire of the child to punish parents indirectly through the use of drugs (or other forms of self-destruction), but also the desire to have parents set limits in order to show that they care. The neglect by parents of stopping the child from doing those things which the child himself deems dangerous is viewed by the child as *indifference*. This situation arises frequently in children whose parents have avoided setting limits—with the excuse to themelves that they are being progressive, liberal parents. The child then must leave messages about his destructive behavior to the parents, as this girl had done, so that the parents are fully aware of what the child is doing. Unfortunately, by the time the child has reached the age of adolescence, where this type of behavior is possible, it is far too late for the parents realistically to put a stop to it. They must often use an auxiliary device in the form of the courts, the police, or a therapist.

The child still views outside help as neglect on the part of parents. The child inevitably maintains the fantasy of the parents' omnipotence. Parents are often the first to know about the child's drug abuse, and it is often at this early stage that the first steps can be taken to make an attempt to limit the child's behavior, despite the loud verbal protestations of the child. If the parents become highly agitated over the child's behavior, however, they are only feeding the child's need to punish the parents so that they will care about him. Many parents of drug abusers are prone to hyperexcitability; their response to these situations tends to perpetuate the child's behavior.

Another common need of a child is to indulge in self-destruction because of a primary masochistic desire that has been repressed to the unconscious. As has been mentioned previously, such children suffer an overwhelming sense of guilt, and much of this guilt is attributable to their intense feelings of rage. The adage "an eye for an eye" is almost always brought into play by the primitive superego. The child believes that he deserves to be punished in a manner similar to what he wishes upon others. In order to alleviate guilt, self-punishment becomes a necessity for the unconscious aspects of the superego.

It is easy, then, to see that self-destructive behavior that is not causing pain for oneself becomes the ultimate compromise for such an individual. Such behavior causes pain in those people whom the child wishes to punish; it proves that these same people care; at the same time, it brings about the punishment which his superego dictates that he deserves. This kind of all encompassing defensive compromise makes it very difficult for the child to give up self-destructive behavior, as such a renunciation would require recognizing three separate components that as yet are unacceptable to him.

It is now clear why the drug-abusing individual is one who suffers from low self-esteem, low self-worth, and the inability to function productively and sexually. He cannot face these feelings in himself; he uses drugs to alleviate these feelings. However, it is also clear that when somebody is heavily involved in drugs he cannot be expected to produce or perform at work or play. His desire for self-destruction is self-fulfilling.

THE DRUG AS AN EXCUSE

The drug abuser often uses the use of drugs as a rationalization for his poor productivity and poor sexual performance. In one case, a young man was able to deny his sexual impotence and blame it on the fact that he was always stoned on barbiturates whenever he attempted sex. Prior to any possible sexual encounter, he would take two barbiturates and would then be able to explain to the girl that he was unable to perform because he was stoned. He knew previously that he would not be able to maintain an erection, but he found this psychological condition intolerable to explain to a girl. As most of the women in his subculture expected sexual relationships with men, he thus avoided ridicule by his friends.

With many of the youngsters who have been unable to abandon their parental values, we find that they have to justify their inability to meet certain standards. They use the excuse that they are stoned on drugs and therefore are not capable of producing what they feel they should. They then tell themselves that they would prefer taking drugs to being productive; lack of productivity becomes a matter of choice. In this manner they do not have to face the fact that they are afraid that no matter how hard they tried they would not be able to succeed. It is always easier to accept the feeling "I do not want to" than "I cannot at all." Drugs are the perfect medium for this type of rationalization. In this subculture it is desirable to take drugs, and so it is acceptable to be unproductive.

The defenses we have discussed in this and preceding chapters are often intertwined and intermixed, sometimes one and sometimes the other taking the forefront. Often, an entire defensive structure will break down and the underlying feelings will emerge; at this time it becomes necessary for the drug abuser to employ one or more of the other defenses. The more intense the feelings or the weaker the ego requiring external support, the more immersed the individual will become in the entire drug subculture. Often a healthier individual will take more drugs, but without immersion in the intricate patterns of drug defense mechanisms. When these mechanisms begin to break down in a therapeutic situation, the underlying psychic pathology begins to emerge. Beneath this complex of defenses lies severe depression, undisguised raw rage, paranoia, delusional thinking, and hallucinations. These symptoms often are hidden beneath a morass of defenses that often are strong enough to keep even the most intense psychotic process repressed. Psychiatric hospitalization is often necessary following a successful withdrawal from the entire drug defensive complex.

It is only by understanding the complexity of possible defenses and the difficulty of therapeutic work that we begin to approach the promise of treatment and the restoration of healthy relationships between the drug user, his family, his peers and society. The following example makes this clear.

P is an 18-year-old white girl from a middle class suburban community. She is currently in a private college where she maintains a dean's list average. On weekends she works as a pianist with professional groups in order to supplement her college allowance from her parents. She began using drugs when she was about 15. In high school she maintained marks in the low nineties. She obtained her drugs from fellow students. Her drugs of choice were marijuana, Quaaludes and barbiturates.

From the time she entered high school she felt "different" from the other students, a feeling that persists today. She had many acquaintances and to all appearances was popular. She was never able, however, to attain a closeness with any of her schoolmates. She never dated, though she is an attractive and enjoyable person. Rather, her interest was towards older men, whom she admired and "loved" from a distance, in a sort of schoolgirl "crush" of the prepubescent. Her main interest in high school and college was her music. Though she is an accomplished pianist, significantly, she never believed in her own ability.

She entered treatment on the advice of a friend because she was depressed and overwhelmed by life. Her first contact was by a telephone call while she was "stoned" on "downs" and feeling very much like killing herself. In the first interview it was apparent that she was a very infantile girl who could not stand pressure, anxiety, frustration, or delayed gratification. She had no ability to set her own internal limits and was constantly looking for a parental figure who would set limits, make decisions and assume control of her life. She had previously found such a person in the form of one of her high school teachers, who had taken a paternal liking to her. It was an asexual relationship by mutual consent. The teacher was no longer available when she was in college, and so she had become frightened, depressed, and lonely. She was not really seeking a therapist, but a replacement for this teacher.

Despite the constant seeking of a parent figure, she was just as consistently rejecting her own. She lived at home with her father, her mother, and her younger brother. The home was rarely tranquil, as arguments and physical fights were commonplace. Both parents are emotionally sensitive people, easily provoked by their daughter's actions. At the same time, P *needs* to provoke. She has been chastised and physically punished on numerous occasions for her provocative behavior. Her relationship with her mother was pathological. Neither mother nor daughter had ever made a complete emotional separation, so neither knew where one began and the other left off. In clinical terms, she and her parents had diffuse ego boundaries and a minor symbiosis. When the daughter experienced success or failure, so did the mother. This relationship was transferred onto the doctor in short order, and was the basis for a number of suicide gestures. The emotional logic behind the gestures was that if she hurt herself the doctor would suffer.

In the course of therapy it became clear that much of her provocative behavior was to satisfy a masochistic need, which often emerged in her relationships with other men. She would consistently place herself in positions with the men in her life so that she would be abused. Accordingly, she often found herself in compromising sexual positions for which she felt guilty later.

The function of drugs in her life was manifold. First, they satisfied the masochistic need by self-abuse. At the same time (since she did not recognize a separation from her mother), it was fantasized that she was getting back at her through self-destruction. It was an attempt to destroy the maternal "introject" which she viewed as evil.

The dynamics of this reaction are based on her ambivalence toward her mother. As she began to mature and attempted to grow independent, she was held back by her mother's intense need not to separate. She developed an overwhelming emotional dependency on her mother, which was at the same time comforting and frightening. During childhood, when the need for independence was not great, she was a "good girl" and very close to her mother. As she grew into adolescence and toward independence, her attempts to break away were thwarted, both by her mother and by her own dependency.

Her intense rage toward her mother was first seen in therapy as an anger at her mother for not letting her break away. As therapy progressed it became clear that there was a second layer of anger (which subsequently was redirected to the therapist and was the cause of suicide gestures). She was angry at the fact that her mother was not always present or always able to comfort her on demand. In order to insure that the mother was affected, she frequently left drugs lying around, either in plain sight or scarcely hidden in the top drawer of her dresser. She inevitably got the desired response of hysteria from her mother and a beating from her father. To drive her point home she went so far as to make sure that her contraceptives were found in a similar manner and with similar results.

The second function of drugs in her life was, of course, to alleviate anxiety in stressful situations. Whether tension came from an exam in school, a date with a man, or a difficult musical performance, she would take a quaalude or barbiturate to relieve it. Unfortunately, and typically, the after effects were worse than the original anxiety. As she had poor ego control and an infantile superego, the drugs destroyed whatever controls she may have had. It was only then that she could allow herself to be placed in masochistic sexual situations.

After the effects of the drugs had worn off, her guilt was excessive. Primitive superegos tend not to function in terms of right and wrong, but in punishment. She developed a need to do penance, and her self-inflicted punishments grew severe. These were accompanied by severe depression and self-degradation, and more drugs. When not under the influence of drugs, however, her morals were rather Victorian; she sincerely believed that sex was something to wait for until one is married.

As seen here, a third function of drugs was to relieve guilt, at the same time allowing her to gratify her masochistic needs. These latent needs could not be satisfied otherwise, as they did provoke guilt. The drugs, finally, supplied a rationalization for her behavior. "I couldn't help myself. I was stoned and did not know what I was doing."

In such situations a narcissistic mother appears to have the best interest of her child at heart. But appearances can be deceiving; the child was no more than an extension of her mother. The mother never allowed the child to grow or develop independently. When time for independence arrived it was a fearful phenomenon. It implied total separation, isolation and alienation from those around her. As she had rarely encountered frustration in childhood it was unacceptable in adolescence. Anxieties in childhood were always thwarted by her mother, so she never felt that she could master anxiety on her own. Her father, rather than being able to be master of the situation, resorted to violence on a par with a temper tantrum. It never helped the situation: the identification with a stable authority figure that is necessary for mature superego formation was lacking.

This case further exemplifies the many pressures at work in an individual with an infantile personality pattern—a common drug abuser. The treatment is difficult because of the early age at which the infantile pattern begins to develop. It is further complicated by the frequency of suicide gestures. Death itself is rarely desired, but it may occur accidentally during an attack on the "introject." If the therapist is unwilling to run this risk there is no chance of improving the youngster's condition. If the patient cannot tolerate the frustration necessary to grow, that fact in itself provokes a suicide gesture.

PURSUIT FOR THE SAKE OF PURSUIT

It becomes an active part of every drug abuser's day to obtain the necessary drugs to get him through the next high. Often this pursuit of the drug in itself becomes an important part of the entire drug syndrome and plays an essential role in the adolescent's protective devices against his own feelings.

Perhaps the most obvious of the gains from the pursuit of a drug is the financial reward. It takes little effort to sell drugs once involved in the drug scene. A drug user can maintain himself in his pseudo-independent state financially without having to run the risk at an enterprise at which he might fail. Through this he develops a false sense of pride and self-esteem. Selling drugs is often looked upon as an accomplishment, much as a little child sees a minimal effort as a maximum success. One highly intelligent drug abuser who supported himself for a number of years in peddling told me that initially he told his parents about it because he sought their approval for his business success, just as he would from getting a hit in a little league game or from playing a musical instrument. He had anticipated that his parents would be as proud of his success as a drug dealer as they were about any other accomplishment.

The pusher also receives recognition from other drug abusers. The more an individual deals, the more respect he gets in this subculture. Other boys look up to him for his financial success and its rewards, in terms of stereos and sports cars. The girls in subgroups will pay him the homage and the admiration another group of girls might pay to a star halfback on the high school football team. A secondary benefit is that frequently young girls will have sexual relations with the drug distributor in order to obtain a supply of drugs. As the self-esteem of the dealer is usually low to begin with, he sees this form of conquest over the female as a symbol of his sexual prowess.

In actuality, therefore, the value system of the average dealer is not that different from many legitimate businessmen in our society. The values are those of money, prestige, and enhanced masculinity. The major difference is that not only is the pusher dealing in illegal merchandise but his self-esteem is so low to begin with that he knows he could not make a success of himself in any area requiring effort or competition.

Not only is the dispensing of the drug an important part of the individual's ego function, but so is the obtaining of the drug. For the hardcore addict or for one who is heavily involved in the abuse of drugs, the pursuit of the drug gives a purpose to a life that would otherwise be meaningless. There develops an entire daily structure revolving about the obtaining of his drug. This structure often becomes so routinized that it can be looked upon as a ritual. It frequently becomes a matter of daily socialization. All concerned gather where drugs are being dispensed, to talk and gossip. Without that part of the day, life would often be so meaningless to the addict as to lead to a depression. In many cases, the ritual of obtaining drugs reaches obsessive/compulsive proportions, and in itself becomes a defense mechanism. There are certain magical connotations in where and when drugs will be received.

B, a codeine addict, remained alone and isolated in his room, until time came to replenish his codeine supply. He first had to obtain a doctor's prescription blank by contacting a friend in the Bronx. His trip to the Bronx was the first time that he used to leave his room in three or four days. This was one of the few human contacts he could afford himself. Once he received the prescription blank, he had to forge it for a codeine cough syrup. He always returned to his room in order to forge the prescription. He never allowed himself to do it at any other location, as he became frightened that if he did, he would be caught: the "magical" element common to an obsessive individual.

He made the rounds of various pharmacies which he knew dispensed the medicine without difficulty, but at a higher price than with a legal prescription. Each pharmacy was visited in turn on a specific day. He never allowed himself to visit a Tuesday pharmacy on a Friday for, once again, he was fearful that this would lead to his downfall.

As with many obsessive/compulsives, there was a chink in his ritualistic armor: he compulsively collected all the empty bottles and stored them in his closet. He intended to lie to his parents by saying that he had a serious cough if they ever stumbled upon his supply. His parents were totally uninvolved and oblivious to the fact that he, at age twenty, was absolutely nonproductive and spent most of the day in his room. They really did not become suspicious until they came upon two or three hundred empty codeine cough syrup bottles in his closet. It was only then that they became aware of his drug abuse and got him to a therapeutic center.

Shortly after entering therapy, it became apparent that he was actually leaving the bottles as a message to his parents to force him to get help. He was fearful of obtaining help on his own because he could not shoulder the responsibility for the decision. By having his parents coerce him into therapy, he took the decision off his shoulders and placed it on theirs. Forcing others to make decisions is a common device of many obsessive/compulsive people to avoid the anxiety of decision-making.

As therapy unfolded over a period of months, more of his magic ritual of obtaining drugs became clear. The ritual of the prescription signing was a way of warding off the evil that he perceived in the police. This evil was a projection of his conscience, combined with his anger, and he recognized in therapy that he desired to be punished for the angry impulses he was repressing with codeine. Since he was fearful of punishment, yet felt he deserved it, he had to set up a magic, compulsive ritual to ward off this reprisal.

In therapy he became aware that the ritual of visiting certain drug stores on certain days did not have the logical basis he had surmised. Rather, it represented part of a "master scheme" that he had concocted in his own mind to prove to himself what a sophisticated drug abuser he really was. He complicated his schemes beyond necessity: he felt so inadequate and so unable to function near the level that he had in high school that he had to delude himself with the fantasy that he was a master criminal and that these were carefully thought out plans which enabled him to avoid detection. If he had visited the drug stores without consideration of the day of the week he would have viewed himself as an ordinary drug addict and would not have been able to support the fantasy of being a master mind.

During his high school years he had been successful academically and socially, and it was not until shortly after graduation that he became involved with codeine. His involvement had originally stemmed from severe anxiety in a job he was unable to hold without the tranquilizing effect of the medicine. However, he had developed a self-image in those high school days, and tried to maintain that image by spending any extra money, after what was spent on drugs, on his clothes. Between his high fashion and his intricate schemes he was able to maintain some form of self-respect, despite being totally nonproductive and completely immersed in the drug world.

The various complicated schemes of the abuser to obtain his drugs often go beyond what is necessary to avoid apprehension by the police. Not infrequently, it is the complexity of the operation that leads to an arrest. Note how these schemes often provide an outlet for frustrated impulses as well as a defense against impulses. All defense mechanisms are in fact a compromise between the drive and the negative response to the drive.

—Part Three—

Healing

Therapy takes many forms, but it must break a vicious circle to succeed.

—11— Sense and Nonsense About Rehabilitation

The most publicized aspect of therapy for the drug abuser is the notorious failure that every discipline has experienced. Whether the individual has been treated by classical psychiatric methods or by newer, more involved techniques found in therapeutic community concepts, the success rate has been remarkably poor. Many psychiatric hospitals have totally abandoned any attempt to treat drug addicts, and are highly reluctant to admit to their wards the serious drug abuser. It has reached a point where it is nearly impossible to have a suicidal or seriously psychotic individual admitted to a psychiatric facility if there is a concurrent addiction problem of heavy drug abuse. Perhaps for this reason it makes the most sense, then, to begin a discussion of rehabilitation by attempting to understand why drug treatment has failed in the past.

The biggest hindrance to the treatment of the drug abuser is the patient himself. The total lack of motivation for therapy found in most individuals involved with drugs makes it nearly impossible to form a therapeutic alliance—no matter what the form of therapy is. Unlike most areas of medicine where an individual can be treated at least *partially* against his will, in any of the psychiatric areas the patient's cooperation is absolutely necessary.

Two basic factors determine the psychiatric patient's motivation. Primarily, the person must feel enough intense pain to make him willing to face the rigors of a therapeutic situation. Secondly, an individual must be willing to change, and see the need to do so. Because the drug abuser's life style is perfectly comfortable to him, and he at least consicously chooses his life style, he does not see the need to change. His symptoms are perfectly ego-syntonic: rather than being disturbing to himself as an individual, he is more likely to be disturbing to the community at large. Because of this defense—of externalizing the symptoms—the drug abuser passes the need for change onto the community around him and refuses to face the need within himself. Nearly all of the abusers that do enter into a therapeutic situation do so because of outside pressure exerted either by parents or the court.

The second major difficulty in forming a therapeutic alliance with the patient springs directly from the first. He sees the therapist as an ally of those people who brought him into therapy in the first place. If the therapist is viewed either as an ally of the court or an ally of the parent, it is very difficult for the youngster to trust the therapist sufficiently to form a relationship with him. Because of his years in the street and his constant evasion of the truth, characteristic of all drug abusers, he finds it impossible to trust any individual—even one who holds out his hand in a gesture of sincere friendship.

Both from a projection of his own insincerity and from having been cheated frequently while out in the street, the drug abuser is extremely wary of any new relationship. In the case of classical psychotherapy, the therapist tends to be an older person and is viewed as being opposed to that behavior to which the patient so tenaciously wishes to cling. If the therapist cannot keep the patient in treatment for a long enough time to establish a relationship *which differs from the one the patient has experienced with his own parents*, he does not stand a chance of gaining the confidence and trust that is necessary for successful treatment.

Even in those treatment centers where therapists tend to be young, and not much older than the patients themselves, an immediate distrust exists. The drug counselor is seen as an authoritarian figure. At the beginning the patient will completely disregard the similarity in age and view him in the same light as one who is much older. Because of the nature of the therapeutic community structure, this view of an authoritarian-in-charge persists. The behavior of the therapist often confirms this image. In the character of the drug abuser there is a prime necessity to prove his own independence and rebel against any individual upon whom he can become dependent. Frequently, when a dependency feeling does become intense, usually after a number of months of being in a therapeutic situation, the patient will drop out of therapy. It is very important for the therapist to bear this reaction in mind, and to be aware of the pulling away by the patient when his dependency on the therapist begins to grow. A close, dependent relationship can develop intense anger towards the therapist as the patient becomes more angry with himself for the dependent feelings which he loathes.

Secondary erotic feelings will also arise in the patient toward the therapist, as a carry-over from the early dependency objects in his parents. As the patient's oral needs are carried into later life, they develop erotic overtones. Such erotic feelings are as taboo toward the therapist as they were to the parents. The constant awareness of these "incestuous" feelings and frequent interpretation of them to the patient are necessary to avoid a sudden rupture in the therapeutic alliance. Otherwise, a patient frequently panics from the intensity of the erotic transference.

Encounter groups typically involve close physical contact, which simulates the dependency-erotic nature of the unconscious wish towards a parent. Heterosexual or homosexual, the patient-therapist involvement quickly leads to a taboo situation. Even the classical psychotherapist, who sees the patient only once or twice a week, finds a dependency relationship developing nearly as rapidly as it does in a therapeutic community.

The patient desires a parental relationship so strongly that he will develop a transference of intense proportions to any person who is willing, even remotely, to assume the parental role. The therapist usually has less control over a patient's behavior than he would have in a therapeutic community. Accordingly, he is more likely to be plagued with absences and lateness to appointments. Frequently he is faced with the youngster who refused to return for his next appointment, without any prior warning. *It is important at this time for the therapist to ignore the usual therapeutic distance and make contact with the patient* in order to reestablish a "parental" linkage and give himself one last chance to interpret the patient's fear of his dependency needs.

The dependency need is a normal step toward rehabilitation. In all therapeutic situations, especially in a therapeutic community, rules and regulations that the patient knows he must follow are his first brush with such needs. As the dependency increases, so do rebellious acts: it is important to be aware that the patient is thereby expressing his fear of closeness to the therapist along with his hatred of the therapeutic situation. Too frequently this fact is neglected in the interpretation of a drug abuser's behavior and his rebellion is seen as one more "game."

Let us assume this hurdle is overcome. Now, no matter how intimately the patient becomes involved in therapy, and no matter how well he is able to accept his dependency feeling there is always the remnant of the self-destructive need that originally drove him into drugs. As he becomes healthier in other aspects, the need to destroy himself arises and begins to sabotage his therapy. Usually the behavioral changes are almost complete, and the individual is wrestling with internalized neurotic conflicts rather than with how he "should behave." He seems to be making great strides forward and both patient and therapist are pleased with his progress. At this time, if his desire to destroy himself has not been adequately dealt with, there can be a sudden reversal.

The patient may show a sudden desire to leave therapy, even to the extent of being willing to go to jail if there is a probationary condition to his therapy; or he may make overt suicide gestures. At this time he may return to the use of drugs for what is generally a brief period; but the intensity of the drug abuse usually far exceeds that which he showed on the street. A break in the therapeutic alliance appears, and the patient begins to slip back into a pattern of self-destruction that is usually more direct than it was when he was in the drug culture.

Such problems are intensified by an often unconscious desire on the part of the parents to sabotage the therapeutic situation—even though it is the parents who bring the child to therapy in the first place. They become unwilling to go along with the therapeutic steps necessary to make their child well. The reason for this paradox is that the character pattern of the child has become a type of behavior on which the family depends to maintain a psychic equilibrium. Any change in one member of the family brings about discord and discomfort to the others. No matter how intensely a parent will profess that he wishes his child to change, he often cannot tolerate a healthier child as he does not know how to cope with the new relationship.

In most therapeutic situations, a demand is placed upon the parents of the child to change along with the child. Usually this involves a more intense involvement with their youngster which they basically do not wish to have. They become resentful when it is thrust upon them. Frequently they have pushed their child into the passive aggressive personality characteristic of the drug abuser *because they cannot tolerate his direct aggression. When he changes and his aggression is externalized towards the parents directly, they cannot tolerate the new situation.*

Many parents unconsciously wish their child to remain infantile and dependent upon them. Becoming healthy, in contrast, requires the child to grow away from the parents. They do not like to see their child forming new relationships, falling in love, moving out of the house and becoming truly independent. They often do all that they can to undermine his growth and maintain him on an infantile level.

In more severe cases, where there is a psychotic family structure, as one member in the family becomes well and loses the psychotic symptomatology, another member in the family becomes "sicker" and his psychosis becomes apparent. Often a parent feels he must maintain his child in a psychotic state in order to keep himself free from his own psychosis. This type of situation requires intensive therapeutic involvement with both parents and child simultaneously. It is usually too frightening, however, to either one or the other to maintain the therapeutic situation over a prolonged period.

The difficulty of such a situation was seen clearly in the case of L.S., a fifteen-year-old "acid head" who was brought to the day-care center. His entire family had been in one or another psychotherapeutic situation for a period of years with little success. On his admission to the center, he was diagnosed as being paranoid-schizophrenic, a condition which clearly existed prior to his involvement with LSD.

From his history, it was clear that as his mother received benefit from previous therapeutic situations, he began to deteriorate. Conversely, as he remained in the center and showed steps towards improvement, beginning to limit his psychosis and to deal with the world on a real basis, his mother's condition began to deteriorate. She was highly resistant to the therapeutic assistance she received at the center from a social worker. Treatment in such circumstances requires a separation of the child from the parent; unfortunately, the psychotic tie was so intense that this was intolerable for both of them.

If parents refuse to undertake treatment for themselves there is very little that can generally be done for the child—unless he can leave the home. No matter how many hours the therapist has contact with him, the child inevitably returns to the home situation that has produced the problem initially. Parents who are generally functioning on a level that they consider acceptable both socially and psychically are hesitant to enter therapy themselves. In essence they are suffering from the same type of disorder as are their children, an ego-syntonic condition in which there is no recognition of the need for their own treatment. *It is absolutely essential for successful treatment to involve the parents*, to make them recognize the need for themselves to change not only in relation to their child's illness, but in relation to themselves and to each other. Their own limitations and conflicts that have influenced the child from the time he was an infant must be altered.

Parents have many ways of sabotaging the therapy of their child. The easiest is to pull him out of therapy on one pretext or another. In private therapy, money is one of the first rationalizations that is used. Suddenly it seems that the parents can no longer afford to send their child to therapy, even though they were fully cognizant of its long-term nature and expense when therapy started. Even if they do not pull their child directly out of the therapeituc situation, they often use money as a weapon to instill guilt in him. The child begins to withdraw on his own.

In most therapeutic community situations, where there is no fee, parents usually find another type of excuse. The tremendous personal involvement that they must put into the community is too tedious and not worth the effort for them. They use this as a rationale to withdraw their child from the center; they will try to find a therapeutic situation that does not require so much time.

A more subtle situation exists when the parents resort to ridicule of the therapist's techniques. One young patient heard Freudian concepts so satirized in the home that every interpretation that was made became an object of scorn. He was thrown into such conflict that he nearly withdrew from therapy, until he began to discuss his feelings towards his parents for ridiculing the therapeutic relationship. In the initial contact with the parents, they had acted in this manner under the guise of humor. If their motivation had not been suspected, this patient would have been lost from therapy.

Another patient in the center had a mother who constantly questioned the concept of the cleaning, housekeeping, and "drudgery" work that is part of the center's concept of mutual cooperation. The patient was chided for doing menial tasks that were supposedly beneath her. She came into serious conflict between her parent's ridicule and what she knew the center demanded of her.

These subtle forms of sabotage are usually the most difficult to deal with because they are not easily made clear to the patient. A patient may continue for months in therapy before being aware of the development of a negative attitude towards the therapeutic situation. It becomes the therapist's role to intercede in the parental sabotage. Often it is necessary to call the parents in, with the child present, to point out to them what they are doing. The discussion must be direct and concrete, leaving no room for doubt in the parents' minds that if they continue this sabotage they will undermine any chance the child has of getting well. In such a direct confrontation the therapist should not be afraid of losing the child from therapy because the child is already virtually gone if the parents continue to undermine the therapy.

Many therapeutic failures can be directly attributed to the rigidity of the therapist. This holds true equally for all of the concepts of therapy. Despite the similarities of the pathological character of each drug abuser, there are marked enough differences between each of them, and between the drug abuser and a classical neurotic or psychotic, to warrant very careful individual approaches to every youngster. In the vast majority of cases, any attempt to involve the individual in the classical analytic mode of treatment will result in failure. This does not, in any sense, mean that one abandons the analytic concept of treatment; it is merely the techniques that must be varied and not the well-known theories and principles of psychotherapy. Classical analytical techniques are based upon frustration of the patient, with the slow development of a transferential relationship that permits the neurotic individual to work out, through feelings toward the therapist, early childhood conflicts. This technique enables the past to be moved up to the present; the individual re-experiences feelings that were present in childhood and are repeated in the adult psyche.

The characterization of the analyst sitting behind the couch, giving little or no response to the patient, who is doing most of the talking and most of the work, bears no resemblance to even an analytically oriented psychiatrist treating a drug abuser. In the treatment of most adolescents the therapist must show himself to be a real person, even if to a minimal degree: the child needs a new figure to identify with, different from those of his early distorted childhood figures. This is especially true in treating the drug abuser. The overwhelming deprivation that is usually felt by him in childhood makes it impossible for him to tolerate the frustration of the remote, uninvolved analyst.

If the therapist cannot adjust to an intermittent giving and frustrating, he will soon lose the patient. Otherwise, all he accomplishes is a recapitulation of early childhood experiences that were intolerable to his patients. Then they want nothing to do with anyone who vaguely resembles the parent figure in their adolescent life. Since the therapist is primarily dealing with a behavioral disorder that makes the patient totally unaware of neurotic conflicts or psychic pain, the first step is to make an attempt to set a limitation on such behavior. If there is no limit-setting, the patient "acts out" away from the office and no therapy can be accomplished. The *method* in which the limits are set is important and will be discussed later.

Many therapists find it difficult to relate comfortably to adolescents. They either try too hard to show the adolescent that they are understanding and very similar to them in ideas, or they do not try hard enough and maintain a stiff and rigid posture which is threatening to their young patient. *Most adolescents do not like a therapist to be like an adolescent*. They do not want the therapist to dress, act, or use the language that they hold to be part of the adolescent world. This is not different from the way that they feel about their parents; they resent their parents trying to dress in an adolescent costume and therefore depriving the child of his individuality.

Most adolescents, though they will initially say otherwise,

are satisfied to call the therapist by his last name, do not really object to his wearing a shirt and tie, and really do not demand that he use four-letter words if he is uncomfortable with them. In fact, some of the more ill adolescents will find adolescent behavior on the part of the therapist an intrusion into their own egos and will become very threatened as to where their ego boundaries end and the therapist's begin. Once they become fearful of the penetration into their own individuality, they are liable to withdraw emotionally from the therapist. They set up an emotional distance which is difficult to bridge, in order to preserve their own integrity. It is usually more for the therapist's peace of mind than it is for the benefit of the adolescent for him to act as if he were part of the younger generation.

On the other hand, there is the therapist who must remain an adult, and always preserve his own imagined integrity in an adult world. He will tend to moralize and berate the adolescent for acting in ways that he and his generation could not accept. To moralize to a drug abuser is the fastest way of losing him from the therapeutic alliance. He must not be judged by adult standards; not only is he not an adult, but this particular type of adolescent is more infantile than normal. The therapist must not fall into the trap of allying himself with the child's parents' values, which is tempting because they are so much more similar to his own. If the child's rebellion is no longer self-destructive or destructive to society, the therapist must learn to accept it. He also must learn to accept that the child might continue to smoke marijuana on occasion, or get drunk on occasion, or actually decide not to go to college immediately following his graduation from high school. The therapist must ally himself at all times with any of the healthier aspects that the adolescent demonstrates even if it is in direct contradiction to his or her own ideology.

Perhaps the most serious threat to the actual development of a healthy adult from an unhealthy adolescent situation is the "countertransference" of the therapist. Countertransference is an unconscious process on the part of the therapist towards the patient, when the patient fulfills the infantile needs that are remnant in the therapist. This is the counterpart in the therapist of transference in the patient, and is a potential saboteur of therapy. It is especially dangerous in the treatment of drug abusers because of the number of therapists who have arisen in this field with no psychiatric background. Never having been in psychotherapy themselves, it is very difficult for them to understand their own infantile needs.

The most frequently seen countertransference difficulty in a drug center is with the therapist who needs the patient to be dependent upon him as he "needs to be needed." He, like the parent who refuses to allow the child to grow, will often transmit the message to his patients that in order to please him they must remain dependent on him. As they change, he will suddenly undermine their progress in order to satisfy his needs.

Another difficulty is with those therapists who need the opposite sex to be attracted to them sexually. They will act in a seductive manner or respond to the seductions of their patients—not for any therapeutic reason, but rather to fulfill some lack in their own ego development. Unfortunately, some people who like to call themselves professionals argue for sexually satisfying their patients in the guise of therapy. Many young counselors, without adequate training, have been exposed to this line of thinking through the general press, and use it as a rationalizatoin for their own needs.

Frequently the therapist with a countertransferential response wants to be a "godlike" figure to his patients. Unlike the man who "needs to be needed," this individual can never be wrong. He will often supersede the decisions of the treatment team or his own supervisor in order to carry out his own needs. This behavior reflects upon the youngster he is treating, who must look to him as a god incapable of error in order to feel pleasing to him.

Countertransference is an unconscious process rarely desired by the therapist. If he is told about his needs, he would prefer not to respond to them. It is not a matter of bad intentions. Those individuals who become therapists consciously to fulfill their own needs are not considered here. It is up to each therapist to look to himself, and accept the recommendations and observations of his fellow workers in order to minimize any countertransference interference with the therapeutic process. Ideally, each therapist who is working in the psychiatric field should himself experience psychotherapy, not only as a learning experience but to further minimize the chances of his infantile needs being played out through his patients.

The least important, but still significant, interference with the therapeutic process is society. Demands are placed upon the youngster because of the social situation in which he exists. The school structure will often place pressure on drug-abusing youngsters because the school was originally for normal behavior. Many schools cannot be flexible enough to deal with the emotionally disturbed, and flexibility is exactly what is needed when there are drug abusers in the school setting.

It is often important for the youngster who is in therapy to continue with school. If he is not in a day care or therapeutic community center, it makes his day meaningful and productive and minimizes the chances of using drugs. The low self-esteem that is inevitably a part of the drug abuser's personality can be enhanced by his being successful in a school situation.

Well-intended but misinformed guidance counselors, teachers, principals, and drug counselors who are in the school will often place pressure upon the drug abuser, *based upon the child's intelligence*, completely neglecting the severe emotional disturbance that limits his ability to achieve. They hold on to the same demands that they would for a child of equal intelligence who is not struggling with emotional conflicts. It is difficult, for some reason, for the educator to recognize the need to lower one's goals for the intellectually superior but emotionally limited.

Most of these youngsters are not yet college prospects, and college should not be held up to them as a primary goal. The therapist often finds himself in conflict with the educator in terms of goals, and the child is once again caught in a conflict between two people, both of whom he would like to please. In many cases a child has left therapy because of the constant demands of parents and school officials that he remain in an academic situation preparing him for immediate entrance into college. The therapist insists that he is not yet ready for this type of a situation and tries to limit his goals to an area in which he could be more easily successful. The outside pressure of the school and parents becomes too much for him. He does not have the fortitude to align himself with the therapist, despite frequent admissions of his awareness of the problem. Emotionally disturbed parents, who are unable to lower their goals for their child, join forces with school personnel against the therapist. With the school on his side, the child has a far better chance of standing up to his parents.

The courts can also play an important role in the success or failure of the therapeutic situation. Because the symptoms of drug abuse bring about actions of an illegal nature, the courts are more often than not involved with treatment. If the court is either too punitive or too lenient, the child will not be able to benefit from therapy. The court must be aware that drug abuse is the symptom of an illness and not, in itself, a crime. To put a youngster in jail or in a shelter without giving him the chance to enter therapy will merely perpetuate the symptomatic situation. The child will emerge in much the same state that he was when he was confined. The only difference will be that, though older and wiser, he will have symptoms of behavioral disorder that are increasingly more antisocial and socially dangerous. He will not limit himself to using drugs or occasionally selling them; the chances are he will enter into a more criminally oriented life style in order to satisfy unconscious neurotic needs.

On the other hand, if the courts are too lenient, as in many recent cases, they do not give the youngster a fair chance to enter therapy. The courts, as has been mentioned before, can be the primary motivating force for the drug abuser to enter therapy and *remain* in therapy when the going gets tough. If the youngster knows that his probation officer will allow him to change from one therapeutic situation to another, on his own whim, that youngster will consistently use this method of avoiding any meaningful therapeutic relationship as soon as the internal conflicts arise.

One young woman, who left one day care center to enter treatment at another, had been previously to three therapeutic centers, gaining nothing from any of them. Her probation officer, having her best interests in mind, did not feel justified putting her into a youth shelter as a method of reinforcing her need to stay in one place. It is easy to empathize with what this probation officer must have been going through. Here was a young women who had no criminal record other than possession of drugs. It was difficult to place her in a youth shelter or even threaten to do so because she was quite pleasant. The probation officer should have used his influence, however, to maintain her in any *one* situation long enough for her to have to work through the painful areas that caused her to leave.

How then is it possible to motivate the drug abuser against his conscious will? It has become clear to those who work with the drug abuser that the primary motivational source must be external to the individual. Something must place him in a frame of mind that is more uncomfortable than the thought of entering into a therapeutic situation. Something must be able to exert pressure on that individual, to force him into a situation that he does not wish to have, and to abandon a condition to which he tenaciously wishes to cling.

For the poorly motivated individual the two primary motivating forces can be the courts and the parents. This is especially true with the drug addict, but is nearly as frequently true with the drug abuser no matter how minimal his drug abuse. Only rarely does the internal conflict become so "ego-distonic" that the individual is motivated on his own.

How can a court or a parent force a sixteen, seventeen, or perhaps a twenty-five-year-old into treatment against his will? Primarily, it must be remembered that, unconsciously at least, the drug abuser wishes to seek help. In many instances he is asking unconsciously for limits to be set from which he cannot escape. There is the desire, on his part, for a caring parental figure to assume the responsibility for his behavior, which he is at least minimally aware is destroying him and those around him. It is therefore necessary to ally oneself with the "part" of the individual that has this awareness.

The courts wield far more power over the adolescent and young adult than do the parents. Frequently, however, a parent can make use of the court through such devices as a PINS petition (People In Need of Supervision), which enables the court to assume the responsibility for a child whose actions are no longer controllable by the parents. The court can take the next step of placing the child on probation, sending him to a treatment center. If the child refuses to go, there is still one or two steps possible before placing the child in jail. There are closed treatment facilities under the authority of the Narcotics Addiction Control Commission. There are therapeutic communities where leaving is extremely difficult. Placing a child in a psychiatric hospital is yet another choice. Finally, if the child is unwilling to accept any of these, a short stay in jail or in a youth shelter sometimes allows them to witness the reality of what they have done and its consequences. They then become willing to seek a method of altering their life style.

If the child is still young enough, or still under the parents' influence, the parents themselves may become the motivating force. Direct pressure toward placing the child in a center is the first method available. If the child is reluctant to go, the parent can withdraw financial support from the youngster if he is of legal age; most youngsters are incapable of caring for themselves, so such anxiety can motivate them into a treatment setting. If this, too, fails then the parents themselves may want to enter into therapy—to try to disengage themselves from the pathological, interpersonal relationship that they have with their child. This can secondarily bring about changes in the child as the parents no longer respond to the child's destructive behavior.

Mrs. H first came to me to discuss her daughter. Her daughter was living in Chicago, where she had been attending college. She had dropped out of school and was living with a boy who was reputed to be a heroin pusher. Her daughter denied using any heroin, but it was highly probable that she was using other drugs. She was working as a waitress and essentially supporting her boyfriend. Though the sexual relationship and the drug abuse were intolerable to the parents, the daughter was perfectly complacent in this new life style and completely unwilling to enter therapy. The mother, being severely distraught, decided to enter therapy on her own.

Initially, most of the discussion centered on the daughter and the daughter's activities. She would frequently ask advice as to how to treat her daughter when she visited, and how she should respond to the boyfriend. Soon she began talking about herself and her relationship with her daughter. It became clear that for the first nineteen years of this girl's life she was an extremely obedient and complacent child, never giving the parents any difficulty or cause for worry. At the same time, she had never gone through the adolescent rebellion that is a normal part of asserting her individuality. Because of a rather rigid upbringing she was fearful of any behavior that might distress her parents. Though she had always appeared to be essentially independent, soon after she arrived at college she began to express anxiety about living away from home.

Her mother had always held her extremely close and did not give her a chance to grow. Because of the mother's own anxieties, she was overprotective of her daughter and did not recognize her need to be different from her mother. As therapy developed, her mother became more involved with herself as an individual and began to separate her daughter's behavior from her own needs. It was clear, through contact that the daughter maintained with her parents, that a large part of her behavior was directed against them in infantile anger for their repressive ways. Every time that the daughter reacted this way, she upset the home, once to the point where her father threw her out, forcing her to return to Chicago before the end of her vacation.

Throughout the therapy, emphasis was placed on helping the mother separate from her daughter. Not having had a good relationship with her own mother, she had placed a lot of importance on a mother-daughter relationship to replace the one that she felt she lacked. She had reversed roles, placing her needs upon her daughter in a manner similar to the way she would have liked to have placed these needs on her own mother. Her daughter, sensing this in anger, used it as a weapon against her mother.

Once the mother began to separate, however, the weapon was no longer useful. Her mother began to accept her daughter for what she was, making visits to Chicago and not becoming overtly upset at her squalid living arrangements. Conversely, when her daughter arrived in New York with her boyfriend, they were accepted into the house and to dinner, despite the parents' loathing of her life style.

As the daughter began to feel more accepted as an individual, she no longer had to prove to herself and to her parents that she was separate from them, and her behavior began to change. She had what amounted to one last fling when she and her boyfriend went abroad for three months. This turned out to be a disastrous trip. Upon their return the daughter recognized that it was no longer necessary to rebel against her parents' values in order to be an individual. She could accept some of the values that she found suitable for her life, and reject others. What most enabled the daughter to make this separation was the ability of her mother to separate and allow her to grow as an individual.

On her return home the girl was suffering from mononucleosis and required a hospital stay in New York. She and her parents began to develop a relationship that they once had; she no longer insisted on dressing in blue jeans and a sloppy shirt in elegant restaurants and once again was able to return to a nonrebellious life style.

The major difference was that now she was not relying on rebellion to prove her individuality. Her mother felt far more comfortable with her daughter as she was, and the relationship blossomed into that of two adult individuals rather than a mother-child, child-mother relationship. The daughter eventually returned to college, this time in Ohio to pursue a career in music, manifesting far less anxiety than had ever been present upon previous separations.

It is probable that at some later date the daughter will require psychotherapy of her own. This time, however, she will not mask it behind a behavioral defense system that makes any approach to therapeutic intervention impossible. The parent, by altering her own psychoneurotic need, had brought about a healthy reaction in her child and a return of her child to a nondestructive life style.

A child can be led to a therapeutic situation but cannot be made to join in a therapeutic alliance. Motivating the child to receive treatment, once he is placed in a therapeutic milieu, becomes the first order of business of the therapist, or in the case of a therapeutic community, of the older members of the community. The entire group of workers allies itself in trying to motivate the new member. Staff and older members work together to try to reach those areas in the new member which are most vulnerable.

The great hunger for object relations that an addict usually has is the major weak spot in his defense against joining the alliance. If from the beginning he sees a concerned and interested group of individuals, then he is more than likely to move in the direction of forming a relationship. Therapy itself can come later. The drug abuser is interested in the basic relationship that you are offering to him.

Offering the individual the chance merely to talk to an interested listener will very often be the first step in obtaining the relationship that you are looking for. Other times it may be necessary to be an active participant, in a conversational manner, talking to the patient about those things that interst him and constantly wondering with him why he is attempting to push you away. You can point out his past loneliness and the type of anxious existence he led while using drugs, and during the conversation bring in the question of why he has the need for a self-destructive mode of action.

It is important at this time to bring him into a keen awareness of the unpleasant aspects of his life style. With many youngsters it is necessary to assume the role of a benevolent parent, similar to the idealized image that the patient has of what he would have liked his parents to be. This requires guiding, directing, listening, and trying always to understand, and, if not understanding, keeping the patient constantly aware of your attempt to understand. Frequent reminders to the patient that you are intersted in his problem help overcome the feelings of being forced into therapy.

In a therapeutic community, when an individual is coerced to attend by his parents or the courts, the very fact of *limiting the behavioral defense mechanisms often brings about enough internalized discomfort to make that individual want to seek help*. He then begins to join the mainstream of the community life and to reach out to the other members for support, guidance, and love. This is the first achievement of the therapeutic process.

To summarize, the key to successful treatment is to keep the patient willingly in therapy. At the early stages there are two processes that may cause a patient to withdraw from therapy. First is his growing awareness of his dependency on the therapist. This would cause the same discomfort as did his dependency on his parents. He will become angry with himself for feeling dependent and then externalize his anger toward the therapist for initiating the dependency need. He will frequently withdraw his feelings from the therapist and leave the program as a defense against the growing dependency. The therapist must interpret for the patient that he is fighting against his own dependency needs in a manner similar to the way he did with his parents. Secondly, the close feelings that will develop toward the therapist will be confused with erotic feelings, causing a withdrawal from this anxiety-producing relationship.

Thirdly, the egosyntonic nature of the illness requires constant interaction between therapist and patient. Alternating between fulfilling the needs and frustrating the needs usually can be the most rewarding.

—12—
One-To-One Psychotherapy

Assuming that the patient is motivated enough to come to the office when required, either because of his internal pain or because of external pressure, then the process of psychotherapy may begin. Ideally he should be seen at a minimum of twice a week, and often psychotherapy is necessary three or even four times a week. Over the early hurdles, especially, which he does not have the ego strength to meet on his own, frequent visits are essential. The patient should be seen in a face-to-face therapeutic position; he is rarely able to tolerate the frustration and feelings of solitude and rejection brought about by the traditional couch techniques. Many patients will have experienced some form of therapy before and frequently will have been members of a therapeutic community. For those people, as well as for those who have never been in therapy previously, it is necessary from the beginning to carefully explain this particular therapeutic process.

Unlike many adult patients or healthier adolescents, youngsters often have so intense a fear of the relationship that it requires an intellectual alliance prior to an emotional one. To cover this fear the youngster has typically developed a wall of skepticism, rejection of help, and a rebellion against all adults. By having him understand in a rational way what is about to take place, much of the fear and rebellion is alleviated. Despite the intellectual beginnings, very soon he is caught in the throes of the transference relationship.

From the onset of therapy a delicate balance must be maintained at all times by the therapist—between how much to feed the child's oral needs, and how much to frustrate them. If the frustration becomes too excessive, it will too closely recapitulate his early childhood experiences, with the result that the patient will soon leave therapy. However, if the therapist feeds his needs too quickly, the patient will continue coming, but will make no progress in therapy.

One of the ways of deciding when to feed and when to frustrate is to watch for transferences. Those feelings towards the therapist that are transferential should not be fed; the feelings arising from the frustration of these needs can be interpreted. As the relationship has a large "reality component" in terms of guidance and direction, those areas that are based on reality can be fed much of the time.

If the therapist believes that the patient *can* handle the situation that he is inquiring about or that has been brought about by his behavior, then frustrating the desire to have the therapist assume a directing role is necessary. In this process, an interpretation must be made to acquaint the patient with his infantile desires. Even feeding the patient's needs must be done through the techniques of interpretation, rather than by merely giving the patient what he wishes. The following brief dialogue exemplifies the combination of feeding and interpretation that is often necessary in the "reality" relationship.

In one case, a seventeen-year-old girl, who is about to graduate from high school, was plagued with the question of whether or not she should enter college in the fall. Most of her experiences were marked with failure until her senior year, when she was able, through the help of therapy, to allay her self-destructive behavior and pull her marks up high enough to enter a community college. Her home situation was unsupportive; her father's business called him away for weeks at a time. Her mother worked and even when she was home there was continual anguish between them.

A strong parental transference developed towards the therapist, which fluctuated between the maternal and paternal components of her early childhood. Her father, despite his absence from the home, had always been the more understanding of the two parents and the one who was closer to the girl as she was growing up. The absences throughout her later childhood and adolescence caused frustration and anger at the neglect of the more responsive parent. As a result she had broken from both of her parents, and was trying to maintain some form of independent existence. At this stage the therapist was soon in a parental transference as a kind and understanding male figure, who was authoritarian to a degree, but benevolent. She wished to rely upon his maturity and experience, as she had relied upon her father when she was a little girl, and she was frightened at the thought of his possible disapproval.

The dialogue started with a question, "Doctor, do you think I should go to college in the fall?" A simple yes or no answer was all the girl looked for superficially. However, if the doctor had merely answered, "Yes, I think it would be a good idea," then anxiety would have arisen in the patient over the fear of displeasing the doctor's wishes or attending school. At this point she might have withdrawn from the therapeutic alliance in an attempt to dispel the anxiety, admitting to herself that she was afraid of displeasing the therapist. A simple no answer would have reconfirmed her own feelings of inadequacy and would have made her assume that the doctor was displeased with her. Because of the nature of the transference at that moment, she would have confused the therapist's response and the father's earlier expectation of her failures. The dialogue, instead, went as follows:

Doctor: How would you feel about going to college?

Patient: I really don't know at this time. I was also thinking of working as a bilingual secretary.

D: What would be the advantage of either one of the situations?

P: If I went out to work right away and then went to college at a later date, I would be earning some money and be more independent from my family. I would like to go and get an apartment and possibly move in with my boyfriend by this summer. However, if I go to college now, the chances are that I will be able to have a better job when I graduate and then I will not have to quit the job in order to return to school.

D: Which seems more preferable to you now?

P: I don't really know. It's hard to tell. Right now I think I would like to go to college, but I do not know what I would like in six months.

D: Are you looking for some sort of guarantee that you feel that I would be able to give you if I made this decision for you?

P: I think I am—I'm always afraid when I have to make a decision about what is going to happen if I think about it for the future.

D: If you choose to go to school now which is what you seem to prefer at the moment, what disadvantages do you foresee?

P: Not too many, really. I can't stand living at home, even though it's gotten much better there lately and I would like to move in with my boyfriend by the summer. However, I think that I'm really afraid that if I go to college it's going to end up like it did in high school and my work won't be too good.

D: Apparently, then, a good part of you is afraid of going to college at this time, and is procrastinating in order to put it off to a later time, when you feel that you might be less frightened.

P: That's something that I have always done in the past. I've always tried to push things off into the distance whenever they seem frightening to me, and Lord knows, enough things in my life have seemed frightening when I set out to do them.

D: Well, as the way you have been handling things in the past have never done too much good for you, perhaps you might want to attempt a new way of handling frightening situations.

P: I guess that would mean my trying to get into college in the fall, even though I'm afraid. I'm afraid of trying something new like this and I've always backed out. I would really like to go to college and get myself a good job. I think I would become sick of being a secretary after a short while.

D: In all probability you would feel much better about yourself if you do go to college than if you go to work because this is basically what you want and you would only be avoiding it out of fear. This year in high school you have been much happier with yourself since you have been doing well and you haven't been cutting classes and missing tests because you have been afraid of them.*

This brief dialogue is an example of how to gratify the patient indirectly while interpreting the anxiety that has developed in new situations. The gratifying takes place in giving the patient

*Since this dialogue the girl did go on to college and was on the Dean's List the first two years.

support in statements like, "You would feel much better about yourself if you did what you wished rather than run away out of fear," rather than patting her on the back and saying "You would be a good girl if you went to college." It is important for the patient to get her support *by being aware of her own feelings*; it is important that *her feeling of doing what she wants pleases the therapist*, rather than her mere desire to get his approval. This distinction places a great deal of reliance on the patient's superego for guidance. The therapist must assist the patient in making his or her own decision about his or her own wishes.

One of the questions that always arises in the therapist's mind is that of confidentiality and privileged information. He is working with youngsters who are still responsible to their parents, and parents often wish to have a current knowledge of what is taking place in the therapeutic situation—frequently they say they deserve to know because they are paying the bills. The other side of the coin is that the patient is a self-destructive individual, whose behavior outside of the office is reprehensible to his parents, and who will tell about incidents that are going on in his daily life which can lead to severe consequences for him.

It should be made clear in the first session what the responsibility of the therapist is in this sensitive area. Parents should be told they are not going to be contacted frequently, and that the therapist is not going to act as a disciplinary assistant to the parents. The role of a judge in family disagreements is often wished for by parents who have lost so much control of their child that they are seeking any help possible to stop the child's behavior. This role should be ruled out.

Secondly, the therapist must make clear to both the patient and his parents that he will not be in contact with the parents without the permission of the patient and, if the parents should call out of their own concern, that the patient will be informed of the conversation. It is important for the patient to know that the therapist is being totally honest with him, and in return expects total honesty from the patient.

During the first interview with the parents and child, the therapist should confirm that he will contact the parents *without* the patient's permission only in those situations which he deems irreversibly dangerous: the patient becomes acutely suicidal, acutely homocidal, addicted to heroin or barbiturates, or is in danger of imminent arrest. Other areas in the patient's existence, though dangerous and often disconcerting to the therapist because of the threat to the well-being of his patient, are usually reversible and can be discontinued and corrected later with therapeutic assistance. These may include abuse of drugs, without danger of addiction, chronic truancy, risk of pregnancy, and running away from home. When these situations do arise, the parents inevitably contact the therapist in an attempt to determine what has happened, and it is up to the therapist to decide how much information he wishes to divulge. He must remember that divulging any confidence without the patient's permission, even though the patient is a minor, is a rupture of the therapeutic alliance. It may also have serious repercussions with the therapy later.

A patient often withholds information that he assumes the therapist considers dangerous. Yet it is more important for the therapist to be kept informed of the patient's behavior than it is for the parents to know what is going on in the therapy session. If the situation reaches such grave proportions that the therapist considers it absolutely necessary for the parents to know something, it is most advisable to inform the parents and the patient in a session together. Then the patient is completely aware of what is happening and does not conjecture that his parents have been told things that he would not want known. Once the patient is aware that the therapist is not conniving with his parents behind his back, he will become more easily accessible in discussing his behavior patterns and his fantasy life. Nevertheless, as a parental transference develops the patient will often withhold information which he thinks will displease the therapist—without having the same concern about displeasing his parents.

It is important for the therapist always to be honest with the patient, to answer honestly those questions that require answering. Yet the therapist must remember the nature of the character disorder that he is treating, and not believe everything the patient tells him at first. Most youngsters who come for treatment are highly manipulative and seductive in their manner; they are easily likeable when they want to be. There is a strong urge on the part of the therapist who has a relationship with one of them to take what they say for truth too readily, and to want to believe that things are going well. Especially at the beginning of therapy, anything the patient says might be considered a deception, a distortion, or an omission—until proved otherwise.

The patient should be made to realize the therapist's skepticism is based on evidence rather than upon speculation. If the patient believes that the therapist mistrusts him only because he was a drug addict, he will begin to distort and fit himself into the mold he perceives the therapist has of him. By listening, the therapist will inevitably find inconsistencies very early, and it is upon these slips that he can base his expression of distrust.

Once the lying and distorting have been brought out into the open, it is the therapist's role to interpret the patient's need to lie and distort. What repercussions of the truth does the patient fear? The interpretations may take the form of, "You seem to feel that you have to lie all the time. Is it because you are afraid that if you tell me the truth I won't like what I hear?" Or, "You are apparently so afraid of others seeing you the way you see yourself that you have to lie to cover it up." If the therapist, in short, avoids moralizing, preaching or reproaching the patient, the deception will begin to diminish within a few months.

Early in the course of therapy, the therapist begins to set limits—again without moralizing, which will only push the patient further away. If limits are not set explicitly and if the patient does not feel a vague disapproval of some of his behavior, the behavior pattern will persist and work against the therapy.

Limits are best set with the aid and defense of intellectualization, that is, *wondering* together with the patient why it is necessary for him to act in an undesirable manner. The feelings and thoughts that take place during his actions are explored, and no antisocial act is allowed to pass unnoticed. Any language in the direction of "I do not like what you did" or "I think that you did something bad" should be scrupulously avoided. Such language merely adds to the patient's facility for acting out when he is angry at the therapist.

By turning everything around to the point of view of the patient, the therapist can bring out of dormancy the healthy aspects of his superego. The right approach is in terms like, "You must have felt pretty awful when you were afraid you were pregnant," or "You seem to be willing to pay a high price in terms of your future for remaining absent from school." Very often the concrete approach of *putting a price value on his activity* helps the patient to see more clearly what he is doing. A reluctant patient might readily respond when asked if the chance of brain damage is really worth the enjoyment of a "high." By letting the patient make the judgment, the therapist remains disengaged from the patient's behavior and cannot become a target for self-destructive attacks.

Similarly, disengagement must be maintained during the patient's successes. When transference is positive, the patient will often act agreeably in order to please the therapist—a throwback to the early childhood experience of trying to please the parent. Yet the patient must be taught the value of self-satisfaction and self-gratification, rather than mere parental pleasing. The therapist must avoid the pitfalls of becoming a target for destructive behavior if the transference becomes *negative*. The therapist must therefore avoid all indication that he is pleased with the patient's actions, *but rather show he is pleased that the patient is pleased*.

A supportive statement sounds something like, "It is good to know that you are so pleased with yourself for passing the course," rather than, "I am pleased that you were able to pass the course." Positive reinforcement of success is an active part of the limit-setting that is necessary. Usually in childhood the patient has experienced merely the stick, but in a therapy situation the stick must be handed over to the patient and the carrot brought into clear view.

A shortcoming that is often seen among therapeutic community counselors is their tendency to take the failures of the patient personally: to become distressed following a school failure, a pregnancy, or a slipping back to drugs. Taking the patient's behavior personally is a sure way of allowing the patient to retaliate when angry. It is recapitulating the parental attitude toward the patient, especially when it was a narcissistic parent whose hopes rose and fell with the success and failure of his child. The therapist must establish a distance at this time, and ask the patient how he felt about the failure. If the patient is blasé, the therapist should begin to *wonder with him* why he is so calm after he hurt himself so badly.

As therapy develops, the need to support the patient will begin to diminish and the need to frustrate will increase. Being able to "titrate" frustration and support in the correct dosage and at the right time is the art of the therapist. If the therapist is "all giving," no internal character changes in the patient can be made. Changes will be superficial and based upon transferential reactions toward the therapist. A patient will often improve to please the therapist rather than change to please himself.

In order to understand the *art of frustration* one must first fully understand the meaning of "transference." Transference involves the redirecting of unconscious infantile feelings that have been repressed since infancy from the primary object—parents, teachers or siblings—to the therapist. It is an unconscious process, without the awareness of the patient. It is the prime tool in allowing the therapist to explore with the patient infantile feelings that would not be present in any other way.

In actuality, transference brings the past into the present, by means of reactions and feelings between patient and therapist. These reactions and feelings are indicated in words, in acting out, and in fantasies and dreams. "Acting out," it should be remembered, is simply behavioral response as a defense against therapy. If transference is not used in therapy, all of the feelings about past events will be in terms of memory; transference causes these feelings to be discussed in terms of the present.

In the case of most neurotics, whose conflicting wishes are based more upon fantasy than upon reality, it is possible to sit back and allow transference to develop and be discussed from the fantasies of the patient. It is further possible to allow the patient to become frustrated, and to express wishes that are being frustrated through his fantasies. With the drug abuser this is not possible. Because of his consistent behavioral defense structure, most of his fantasies are acted upon, and his fantasy life remains impoverished.

For example, a young female drug abuser who has strong oedipal wishes towards her father is likely to go out, get stoned, and have intercourse with some older man. If this takes place while she is in therapy, one can assume she has unconscious fantasies of desiring her father. Since her behavior so closely approximates her wishes, however, her fantasies are very difficult to bring to the surface. Further, with the fulfillment of these fantasies through her actions, she is liable to withdraw from therapy. On the other hand, if the therapist assumes some of the functions of the paternal role that the patient seeks, while frustrating any attempt for physical closeness, either in terms of affection or sexuality, then her sexual and physical fantasies will begin to develop towards the therapist.

Now the chances of her seeking support from men outside of the therapeutic situation are diminished: the intense infantile nature of her desire for paternalism is as important to her as sexuality. If, on the other hand the therapist feeds these physical needs, the patient will never be aware of the intensity of her rage at the frustration of childhood desires. She will constantly make demands on the therapist to fulfill her physical desire. The unacceptability of sexual feelings toward the therapist will then lead to discontinuing the therapy.

With most drug abusers, sexual feelings mean far more than sexual intercourse. Coddling, fondling, stroking, hand holding, or kissing good-bye may, one or all, be greatly desired by the patient, who is unconsciously angry about not receiving them. In order to begin to bring this unconscious anger into conscious light, it is necessary to frustrate these desires and talk about the angry fantasies that develop towards the therapist. When the patient is clear in her own mind as to how angry she is at the therapist because of frustration, she can then be told that her basic anger is coming from a wish for a father because of the absence of her own.

The specific situation of frustration of oedipal wishes of a female patient can be generalized to the frustration of any infantile wishes that the patient has maintained. The frustration of maternal transferential wishes for support and nurturing follow a similar pattern. Likewise, the more mature, conflictual transference of a patient's struggle between dependence and independence has the same structure. As the dependence-independence conflict is so intense in the drug abuser, we will concentrate on an example of how it is brought into awareness in a therapeutic situation.

A sixteen-year-old patient of mine had been proclaiming her independence for the previous seven years. She was involved with groups outside of her home, smoking marijuana to excess, abandoning her parents' desires that she go to college, and seeking boyfriends who were unacceptable to her parents. After several months of twice-a-week therapy, she appeared in the office one day and brazenly proclaimed that this was going to be her last session. Nothing that I said seemed to convince her otherwise. When asked why she had suddenly decided to leave therapy when we both agreed that she was making progress, she said that she was tired of not being able to be independent and she did not like how dependent she was on me.

I reminded her that I only saw her for two fifty-minute sessions per week. How could I then be depriving her of her independence when she had the rest of the week to be on her own? She pondered this for a moment and said that whenever she is about to make any move I come to her mind and she becomes afraid of how I would judge her action. I then asked her if at any time I had passed a judgment on any of her behavior, and she promptly replied that I had not. I then made the interpretation to her that she had always wanted me to assume a more active role in guiding and directing her, but that this desire was so contrary to the way she liked to view herself that she could not accept it. She agreed that for some time she had been aware of fantasies that I would come and take care of her; when she had been away the previous summer, she recalled fantasies, which she had since repressed, that I was there to help her make her decisions.

She did not leave therapy as she had originally threatened, but over the ensuing weeks we talked more and more about her desire to be dependent upon an adult figure. From this it was easy to make the transference interpretation that she desired her parents to be more active in guiding her life. She had come from a divorced home, which had forced her mother to work. She had felt for years the absence of any guiding hand and, because of her inability to admit to herself that she had wanted such guidance, she was also unable to recognize how angry she had been at both parents for not being at her side when she needed them. As soon as she came to this realization, therapy progressed smoothly and she was able to act in ways that were completely independent from her parents' desires.

She was now able to understand what I meant when I pointed out to her that to say no to her parents all the time is to be equally dependent upon her parents as to say yes all the time. She began to see that she always said no, no matter what the parents' request. Within a relatively short time she was able to pursue a productive course, borrowing some ideas from her parents, seeking advice from them when she felt she needed it, but generally being able to make decisions on her own.

The therapist's constant repetition of the interpretation, "You want me to take care of you and you hate this desire," in as many forms as is applicable to the circumstances, is one of the keys in breaking through the drug abuser's defensive structure. Conversely, consistently taking care of the patient—by answering all questions, making decisions, and giving unquestioning support—never allows the patient to be fully aware of how badly he wants to be taken care of. He then is never allowed to come to grips with his intense rage at not being cared for, which is more detrimental. Further, because of the massive deprivations the patient has suffered in childhood and because of the intensity of his dependency needs, trying to fulfill them is like attempting to fill the Grand Canyon with a teaspoon.

In most cases there is little risk in losing the patient by frustrating his transferential needs. Unlike the psychotic, the young drug abuser has a strong transference relationship, which is not easily broken or transferred to another therapist. Similarly, the infant who has already differentiated his mother from other adults in his world therefore clings more tenaciously to the mother figure. The psychotic has not yet made the differentiation between his mother and any other adult who cares for him.

The degree of transference often is so intense that in itself it can become a hindrance to therapy. If the therapist allows it to develop to its full degree without persistent interpretations to diminish its intensity, the ambivalence that was fit for the parent will resurge and block the positive alliance with the therapeutic situation. So the therapist must be ready to interpret, "You feel towards me as if I were your mother," or "You are responding to me just as if I were like your mother." He can thereby keep the intensity of the transference down to a point at which it is a useful tool, not losing its beneficial aspects.

The direction of transference is usually maternal, but, as our examples have shown, there are frequent times when the main thrust of a patient's feelings will be toward his or her father.

In those family structures, particularly where the children turn to the father because of a physically or emotionally absent mother, the father assumes the role of the maternal figure. This role is then carried out in the transference relationship; it is extremely confusing because of its overriding erotic nature.

Depending on the sex of the patient, his or her feelings toward the therapist can become either homosexual or heterosexual, both of which are equally taboo for the patient. The therapist is not looked upon as an ordinary sexual object by patients of the opposite sex, but as a transferential object from the earlier person, to whom any sexual feelings would have been incestuous. Furthermore, heterosexual erotic transferences are usually based upon pre-oedipal levels of development, far more attuned to the desire for dependency than the desire for genital gratification.

The erotic component of the transference must therefore be interpreted early; yet not until the therapy has developed to a point where the patient is sufficiently healthy to handle these highly charged feelings can it be allowed to develop to any intensity. If it is allowed to develop at the beginning of therapy, the patient will act out away from the office and repress his feelings in that manner. It may be the first and irreparable step in a rupture of the therapeutic situation. Later in therapy, when conflicts are more internalized and the patient's ego strength has been adequately supported so that he is less fearful of intense, taboo feelings, the erotic transference can develop until it becomes a resistance, and requires interpretation. From the start, an interpretation of the erotic nature of the transference should be made to help the patient to differentiate his sexual confusion between genital and oral needs.

For most drug abusers, orgasm is not the primary function of their sexuality. In a broad sense, the feeling of closeness and the symbolic or literal filling of a cavity are far more important. When sexual activity is broken down into its various components by the therapist and the patient, discussing his or her feelings about each step in foreplay, intercourse and post-intercourse, the patient becomes aware of what he really seeks.

A twenty-one-year-old woman, an ex-drug abuser and ex-addict, who had lived on her own for two years prior to therapy, became highly distressed one day during a session in my office.

She said she felt like running out of the office because she was so uncomfortable with her feelings towards me. When I asked her what these feelings were, she said she felt like "balling." I asked her to describe the exact nature of her sexual fantasy. Rather than wishing sexual encounter, she really had the desire just to come across the room and sit on my lap, to have her hair stroked and to be cuddled—and to be reassured that her decision to go back to college was a good one.

It became clear that this young woman, despite frequent sexual activity, had never had an orgasm during intercourse and could only be gratified through cunnilingus. She found far more satisfaction in oral sex than she ever did in genital sex. Her background was revealing: her father had died when she was thirteen years old and, prior to that time, he had been the person in whom she mostly confided. She would often sit on his lap and have her hair stroked as she told him about what was troubling her. He was kind, understanding, and helpful with her problems. It was just this kind of response that she was looking for from the various men in her life: intercourse was the closest thing that she could accept as an adult that would bring a similar closeness and gratification.

Until the episode in my office, when her erotic transference feelings became intense and were closely scrutinized, she had never been able to accept that she really felt a tremendous loss at her father's death. Shortly after his death she became involved with drugs, she was truant from school, and developed an entirely different life style. It was this behavioral defense structure that enabled her to hide from herself feelings with which she could not cope regarding the loss of her father. The transferential interpretation allowed her to recapture those feelings. Furthermore, she was able to put an end to her sexual acting out, as she realized her promiscuous behavior would never gratify those wishes, and that the only way to gratify them would be in a true love relationship.

If the erotic transference had been allowed to develop—as would have been necessary in treating a neurotic patient—this girl would have maintained a high level of sexuality away from the office, would have withdrawn from the therapeutic situation because of her transferential incestuous feelings, and probably would have returned to her former way of life.

Transference, then, can either be an invaluable therapeutic tool or, if misused, can be the cause of a premature termination. It is the transference *wishes* that must be frustrated in therapy. The therapist must constantly keep in mind that these wishes *are* merely transferential. If this fact is forgotten, it can be flattering to the therapist to think that the patient has made him so important. The therapist can become more entangled in the web of trying to satisfy his patient's needs. He can make the irrevocable error of thinking that he is the one and only person who can give the patient everything he needs, so he sets about to do so. Eventually, because of the infantile nature of the relationship, the therapist will eventually have to fall short of the patient's expectations. Then the patient will make the break from therapy, as it is now too late for any transference interpretations.

It will be beneficial here, after these typical case histories and therapeutic analyses, to go step by step through a generalized therapeutic program. By no means can this order be followed with each individual, nor is this sequence irreversible. Once a certain stage is passed, it can be returned to because previously used defenses inevitably arise during new periods of stress and confrontation. It is, however, something a therapist and anyone involved in therapy can keep in mind as he passes through various stages of work with an individual to keep a check on how he feels therapy is progressing. The unspoken assumption in all of this is that any parent or other adult interested in drug abuse cannot help but benefit from knowing the rationale behind enlightened therapy.

At the beginning of therapy, emphasis must be placed on the behavior of the patient. The therapist helps the patient to intellectualize what he has previously only acted upon. The therapist interprets the patient's behavior in terms of the self-destructive needs and his simultaneous desire to increase the worry and concern of his parents. The therapist also uses limit-setting techniques that were discussed earlier in the chapter. In short, the patient must begin to cut down on his drug taking or any other form of rebellious activity before any progress can be expected.

After the behavior has become limited, which may take upwards of a year in one-to-one psychotherapy, a depression begins to emerge. Once this depression becomes apparent and persists over a period of time, it is then that the patient is beginning to internalize that which he has previously only externalized. It is then, too, that the defense (or use) of intellectualization begins to be a hindrance to therapy rather than an asset. The therapist must interpret this and other defenses standing between the patient and his unconscious.

Depending upon the material that the patient presents in each individual session, various themes in his behavior emerge and can be worked with interchangeably. These may be his anger at his parents, guilt over his anger with subsequent anger at himself and ensuing self-destructive behavior, or a dependency-independency conflict. It is impossible to tell, without knowing an individual patient, which of these three affectual situations will become predominant.

Usually, anger with parents comes readily to the fore because it is easy for the patient to see the running battle that he has been having with them over the past few years. The patient is never really able to understand to the fullest this anger with the parents until he is willing to accept his desires for dependency on them. Otherwise, he will keep trying to rationalize his anger and to bring about situations with his parents in order to justify his attacks on them.

Guilt and self-destruction are also usually more accessible to the conscious mind than are dependency feelings. The patient soon becomes aware that he must have forced that anger out of his consciousness for a reason, and once this fact becomes apparent in its fullest extent the guilt will rise simultaneously. The recognition of his guilt because of his anger usually is the key which allows the recognition that his parents have remained meaningful people to him throughout his adolescent years despite his vigorous denials. It is at this time that he can become fully aware of his dependency needs and that he begins to understand, on an affectual level, his anger and his guilt. When the patient is fully aware of his previously repressed feelings, it is necessary to deal with his feelings of inadequacy, confusion about sexuality, or fear of his impulsive behavior.

There are inevitably many causes for the symptoms of his depressions. They are usually based upon his own sense of guilt, with subsequent feelings of worthlessness, which often stem from the fantasy that his parents would not respond to him the way he wished because he was not worth their time and effort. The patient will always tend to blame himself for his parents' reactions because the infantile fantasy about his parents is that they are perfect. He feels unlovable and begins to search for reasons why his parents would neglect him. He blames his anger on those he sees as family rivals—father, brothers, and sisters—as a cause for being disliked. He further blames sexual feelings, which he feels are taboo.

At this point in therapy the patient often begins to withhold some of his feelings from himself and from the therapist, as he is afraid of meeting with the therapist's disapproval. It is here that the therapist, by interpreting the fear of being unacceptable as a transference, begins to make the patient aware of his own projections as to why he could not be loved. This is not a rapid process in therapy; once again, a better part of the year is generally spent in dealing with the patient's feelings of inadequacy and unlovability. It is because of the overwhelming fear that he cannot be loved, however, that he withdraws from one-to-one relationships and drifts into group situations. When these feelings are worked through, love relationships begin to develop, and with the fulfillment of a love relationship the early frustrations and subsequent anger diminish.

It is now that the feelings that he initially felt for his parents make actual changes, and his libido is redirected towards people of a more appropriate age. His first love relationships may well be of a neurotic nature, where the individual is attempting to seek gratification of his infantile needs. But he now also becomes more responsive to transferential interpretations. The therapist at once must be aware of the possibility of the patient dropping out of therapy prematurely. He has found a new relationship, he is no longer depressed, and he is no longer acting out in a self-destructive manner. He may have a strong dependency attachment to the love object, incomplete sexual gratification, and a consequent fear of abandonment. So it becomes necessary to remind the patient of these internalized conflicts, for he is happier now by far than in the situation he has just come from.

His relationship with the therapist at this time is filled with ambivalence. Intense dependency is coupled with a strong dislike of the therapist *because of the dependency upon him*; there is a consistent fear of displeasing the therapist and ever-increasing erotic feelings based on the dependency and, in some situations, on newly found genital desires. So it is now necessary to begin to deal with the sexual guilt that the patient is experiencing, separating for him the oral wishes towards the parental figure and the genital wishes toward the love object. As his first love object is so closely related to a parental figure, satisfying sex drives with that person is regarded as taboo. Secondly, the ambivalence about being dependent on that individual is so great that the unresolved anger and aggression cause a strong sado-masochistic component in the new sexual relationship.

Simultaneous with positive feelings towards the new love object, similar feelings are present for the therapist, as a transference. It is through transferential feelings that the patient's dependency desires and subsequent anger are made clear to him. As the dependency begins to diminish, his sexuality takes on a more genital nature and becomes separted from the early oral desires towards his parents. Accordingly, he finds greater sexual satisfaction and fulfillment. Consequently, as the dependency and anger diminish, there is a diminution of guilt and sado-masochistic needs.

The patient no longer needs to wrestle with the other individual to determine who controls his integrity. As he begins to feel more independent, he is aware that he is in control of himself and that the love object is not a threat to his new-found self-assertiveness. With the further diminution of his sado-masochistic and dependency needs, the patient will generally find a new love object who does not require this initial neurotic complex to gratify his own needs. The second or third person with whom the patient becomes involved leads to a healthier relationship based on more mature genital love.

It is also about this time that the patient can begin to be weaned away from the therapist. He is becoming more independent and feeling more in control of his life, with less fear of his decisions and their consequences and a decreased need for guarantees. Discussion of termination of therapy should begin between six months and a year before the actual date. At first, intense fears existing without the therapist will arise along with many of the symptoms that were seen earlier in therapy, such as acting out behavior. The patient is, in actuality, recalling earlier defense mechanisms that helped deal with the same anxiety over parental loss that he is now experiencing in losing the therapist. Constant interpretations must be made as to how these defense mechanisms are being called into play again, and how they reflect those of his infantile years regarding his own parents.

Feelings of rejection will increase as termination approaches, and the patient will make an attempt to employ every method he can to prolong the therapeutic situation. Between two and three months prior to termination a definite date should be set. It is generally not necessary to wean a patient away from therapy with ever-decreasing visits to the therapist. If this is felt to be necessary, it is because the therapy has not been successful in resolving the patient's dependency needs. Rather than considering a gradual reduction in hours, it makes more sense to continue therapy until the termination can be ended completely.

Often it is necessary to warn a patient ahead of time that he should be aware of a temptation to return to the symptoms that brought him to therapy in the first place. A warning is often useful that his symptoms might persist for a short time following termination. He can be reassured, if necessary, that these symptoms will disappear on their own as he begins to adjust his life to his new-found independence. Towards the end of therapy, when the patient knows the termination date and knows that whatever situation may arise the termination is inevitable, many of the positive feelings that he had towards the therapist will be displaced to people outside the therapeutic situation. Though these relationships may not be the healthiest initially, he will soon develop associations away from the therapist that gratify his adult needs.

No matter how successfully individual therapy may have progressed, it can be completely undone when the youngster returns to his home environment and his parents are unable to tolerate the changes that have taken place. It is often necessary at this point or far earlier, to encourage the parents to enter into a therapeutic situation on their own. It is usually not advisable for the therapist who is treating the child to be the therapist also for his parents. The child can become confused as to where the therapist's loyalty lies. If he is still in therapy he will very often withhold much of the information which he might be afraid would get back to his parents.

It is sometimes advisable to refer each parent to individual help on his or her own, but more frequently it is useful for the parents to be seen together, concentrating less upon their own psychopathy than on their relationship with their child. A further benefit from their therapy is the informational gain they experience, without having to question their child's therapist and risk upsetting the confidentiality of their relationship. Many parents who cannot accept the idea of going into therapy for themselves can accept it when the rationale is placed upon the need of their child. Not infrequently, this inducement to entering therapy for their child's sake will eventually lead them to seek further needed therapeutic help for themselves.

—13— Group Therapy

Many therapeutic methods involving group therapy, which would have seemed outlandish not many years ago, are now accepted as useful tools in specific situations. They are, in fact, in vogue.

Basically, there are two different forms of group experience. First there is the classic analytic model, which requires a distance between group members and the therapist. In these groups, the therapist does not take part in the discussion, except to interject interpretations and to diminish resistances. The degree to which he performs these tasks varies, by necessity, with the type of group he has in treatment, but, basically, he does not share his own experiences, feelings, and problems with the other members of the group. The second type of group is one in which the therapist takes a far more active role. They include encounter groups, T groups, sensitivity groups, and marathon groups. Here it is the therapist's role to encourage the participants to express their feelings, to share with them his own thoughts, feelings, problems and ideas and the ways he has handled them himself.

The "active" group will often encourage behavioral interaction among its members, with limits set by a group leader and by the type of group that is being run. Behavior can vary all the way from the members hugging one another to show their feelings of warmth and giving, to nude encounter sessions, where some forms of sexual acting out and whole body touching are encouraged. Unfortunately, many of the latter groups are run by therapists who are inadequately trained and motivated. Often the group becomes a medium for the therapist to act out his own frustrations with the help of his patients.

In the treatment of the drug abuser, groups with excessive behavioral expression should not be encouraged. The drug abuser uses so many behavioral defenses initially that behavioral acting out in the group intensifies these defenses and keeps his conflicts and feelings beneath the surface. He finds the group a means of physical expression that is all too often merely a substitute for the behavioral defenses that he has employed prior to coming into therapy.

The first type of group therapy has a far better chance of success. The analytic group is usually composed of eight members, a therapist, and sometimes a co-therapist. It is usually best to have an equal number of both sexes to allow and encourage transferential relationships which often depend upon sexual differences. The group sits and, in the case of adolescents, often sprawls, squats, lies on the floor, or takes any comfortable position. The adolescent is often inhibited by the rigid formality of merely sitting in a chair; the ability to relax in posture facilitates the ability to relax in feelings. If a member of the group refuses to relax physically and holds himself in a rigid postural position, a specific psychopathy is often at work. The therapist might be able to interpret to him the fears and the rigidity that is necessary in his life for the stability of his ego.

One of the advantages of treating the adolescent in a group is that he relies on group pressures and experiences to grow. A youngster may be highly resistant to change when he views a single therapist as a representative of the adult world and an extension of his parents. He finds it harder to deny the sincerity of a group of his peers as they point out to him how he is defending himself.

The interaction of adolescents in a group often allows them to make interpretations to each other in their own language; language which would never be accepted from the therapist. Secondly, the mere physical presence of others dilutes the transference to the therapist, so that the situation is less threatening for the individual. Since many adolescents are struggling with dependency needs that they themselves have to deny, they find that the group enables them to spread their dependency among more acceptable people than parent surrogates. They are accustomed to using a group outside of the therapeutic situation for this purpose, so they can quickly transfer their dependency to the therapeutic group. Members of the group offer a substitute for the initial support that the therapist has to give in individual therapy. Because of this external support, the therapist remains a more obscure figure, and less "reality" plays a role in transference to him.

Not long after a group is formed, ambivalent feelings of the patient toward other members become apparent. On the one hand, there is a great deal of support felt from them, as in a large family; simultaneously, as in a family, each of the "siblings" or group members is vying for the attention of the therapist. This ambivalence towards other members of the group calls up feelings of anger which have often been felt towards family members and which have since been repressed and are lying dormant.

Patients should be aware of how they can have opposing feelings simultaneously. The mechanism of ambivalence becomes clear to them as they experience contradictory feelings towards their fellow group members. At the same time, one member of the group will often become predominant, taking on a role similar to that of the therapist. Other members fall into various situations which are analogous to how they related in groups outside the therapeutic situation. Feelings expressed in the group are more easily acceptable as group members are viewed as peers and not as authority figures.

It is not until the group has been in progress for several months that members begin to feel comfortable in expressing their feelings towards the therapist. These are often outgrowths of feelings toward other members. It is the role of the therapist to interpret such transferences as he sees them unfold and to eliminate resistances to communication between various members of the group.

With drug abusers, however, this ideal situation is rarely in evidence. Because of the inhibited nature of their basic character, and the overwhelming insecurity of expressing their feelings, drug abusers must be approached more actively than psychoneurotic individuals. The therapist will often have to encourage the individuals to talk rather than allow the long periods of silence which are acceptable in an analytic group. Because of the easily frustrated nature of the people with whom he is dealing, long silences breed such overwhelming anxiety that the group begins to disintegrate. Various members withdraw into themselves, in an almost autistic manner, and there is a perpetuation of silence beyond the group. After a while, instead of being beneficial in developing transference, group therapy becomes a liability.

Expectation soon develops towards the therapist and towards other group members. Predominantly, the expectation towards the therapist seems to be an infantile parental transference, based upon the early infantile fantasy of the parents' omnipotence. The drug abuser is basically an orally dependent individual. He anticipates the ability of the therapist to answer all questions, to be continually present for him, and to be able to heal his ills with the facility of a mother for her child.

As the therapist disappoints various members of the group by not being able to live up to these expectations, their feelings intensify—leading to the type of frustration and anger that was felt in early childhood towards a neglectful parent. Along with this, there is an increase in angry feelings towards other group members, who begin to be viewed as rivals for the therapist's attention. This "sibling" rivalry grows, and so does the intensity of transference. Petty fights and arguments break out among group members.

Now the therapist must intervene with interpretations of these reactions. Through this transferential interplay, group members begin to recognize their own infantile hostility towards members of their own family. It is an exquisite reenactment of infantile family life and the subsequent desires that develop in that life. For, along with anger, group members are able to perceive an attachment and even a love for other members as they begin to receive support from them.

As their insight develops into the unrealistic expectations and the infantile nature of their own rage, adolescents in a family become able to cope with these feelings in dealing with peers and eventually with parents. They begin to recognize that much of their anger stems from unrealistic expectation of the parent. They can then accept the parent as a human being rather than as some fanciful, omnipotent person. When this happens, their rage subsides and a different relationship forms among siblings—or, in the case of a group—among group members. Now, however, the similarity to sibling transference diminishes, and there is a more unified attempt by group members to assist one another without the resentment that was seen in the former stages. Each group member, then essentially becomes part therapist and part member in the group, and the role of the therapist begins to diminish. Through the group technique, each member learns the resistances of other members and begins to make interpretations on his own.

Several things differentiate a group of drug abusers from other therapy groups. First, because he has relied on a group for support in the past rather than on an individual relationship, the drug abuser finds it easier to relate in a group setting. Secondly, the drug abuser is able to relax his characteristic defenses of resistance and rebellion towards authority in a group process. Third, the therapist must take a more active role and must be seen as more of a real person at the beginning of a group session so as to avoid the intense frustration which a drug abuser's infantile nature cannot tolerate. Fourth, the fact that there are other members of the group to supply support for each individual allows the therapist to be less active than in the case of individual therapy. Fifth, the familylike structure of the therapist, co-therapist, and "siblings" presents an unusually clear model for the drug abuser as he relives his infantile fantasies and feelings. Finally, the peer group will often be more aware than the therapist of an individual's lies, avoidances, and manipulations so commonly seen in drug users.

A significant stress situation in the group therapy process is the loss of a member of the group. Because of the acting-out nature of their disorder, it is a common occurrence for members of the group to drop out quite unexpectedly. This will inevitably cause an upsetting reaction to the others, who experience the loss as a personal affront. Much of the guilt surfaces from an overwhelming superego, and each member blames himself in turn for the loss of his peer. These fanciful feelings once again become fuel for the therapist to help the patient gain insight into his unrealistic expectations of himself, and into the subsequent guilt that is produced from not living up to these expectations.

When a new member enters to replace the absentee, the group begins to band together, assuming an intolerant stance towards the intruder. He is seen as an outsider, who is vying for the attention of the therapist. Their resentment is usually expressed in a subtle way, to avoid the anticipated antagonistic response of the therapist towards open anger. If early interpretation of this antagonism is not made, it may carry over for a period of weeks and may become disruptive to the group and detrimental to the new member. Most drug abusers do not have the ego strength to tolerate rejection from older members of the group. This is a critical time to make the new member feel welcome so he does not bolt the group prematurely. It must be remembered that, with the drug abuser, *anticipated* rejection is an extremely frequent characteristic, so when the fantasy is fulfilled in actuality the anxiety level increases to an intolerable level.

The ultimate rejection that is experienced by a group member is caused by the absence of the therapist. Unavoidable absences, which cannot be discussed prior to the therapist's leaving, bring about fear and anxiety along with the reaction that the expected rejection has finally occurred. Anger towards the therapist for his absence will usually remain beneath the surface, in order to avoid any conflict with his parent figure. It is often up to the therapist, upon his return, to bring out these feelings by taking a more active role.

When the therapist knows that he will be away from the group, such as for vacations or meetings, he should discuss the absences with the group two or three sessions prior to his leaving. Many group members have difficulty in dealing with anticipated feelings; because of their infantile character, the only feelings that they easily come in touch with are those that are taking place at the moment. When dealing with an especially regressed group, it may be necessary for the therapist to anticipate the feelings of members for him, and to express to the group his own expectation that they may have feelings of rejection, loss, or even anger. At this time, the therapist will be barraged with rationalizations that no one will be angry, such as, "You deserve a vacation—we have no right to expect you to remain with us during your vacation time." These intellectualizations clear the way for feelings to emerge during sessions he misses.

With the adolescent, it is often useful to allow the group to continue in the therapist's absence. The dynamics of the group should continue, and many of the feelings about the absence of the therapist should be shared to make them less threatening. When the therapist returns, he will generally find that progress has taken place in the patient's ability to express his feelings.

A highly useful type of group in treating the drug abuser is the inter-generational, or actual *family* group, made up of a single family or a number of families together. It may include siblings of the patient, depending upon the intimacy and dynamics of the structure of the family. Such a group should vary from structured to unstructured, depending upon the pathology of the family. In some instances, the unacceptable impulses of the patient are so intense that the anxiety they provoke while the family is present may lead to the patient's premature dropping out. This is especially true of more psychotic individuals, whose id reactions are right beneath the conscious level and so are feared to go out of control.

In those individuals with a more highly structured defense system, such as the obsessive/compulsive, families can be allowed to play out verbally many of the feelings that have rent them during the years of the adolescent's withdrawal. The adolescent can separate from his family on a healthy level. Not infrequently, the parents can then allow the adolescent to separate, when previously they have been clinging to him for fear that their child would meet a horrible end if left out of their control. As the dependency-independency conflict is so acute in each drug abuser, this attempt to assist with a healthy separation is necessary. Families should be carefully screened as to their pathology; those with a blatant psychotic interplay may benefit from interaction with another psychotic family. But, if the therapist has too many psychotic families in one group, he will find that no valuable therapy can actually be sustained as the psychotic process becomes too intense and each individual feeds off the other. It is sometimes helpful to place those people with florid id impulses in the same group with an obsessive/compulsive family, so that the high emotional level of the former brings out some of the feelings in the latter, and some of the intense structure of the latter begins to confine the psychotic impulses of the former.

If a family group or an individual family is seen in conjunction with *group* therapy for an *individual* patient, it is usually beneficial to use separate therapists for each. The transference towards the therapist by the child becomes too confusing—when he sees him in the dual role of both family therapist and individual therapist. This procedure is contrary to the approach to a situation that would arise if *individual* therapy is being used in conjunction with group therapy. Here it is usually beneficial for the same therapist to conduct both and in this manner to allow the transference to become more intense. Certain interpersonal reactions that are seen in the group can often be more clearly observed when the individual is alone. Frequently, situations that are brought up in individual sessions are reenacted for the patient more clearly in a group setting. If two therapists are used, however, they will frequently find themselves working at cross purposes, with the patient playing one against the other—as is frequently done at home with parents. The patient is often able to "get away" with this manipulation for too long a period.

Group therapy plus individual therapy is also highly beneficial to the drug abuser because it not only gives him the support of the peer group, but also forces him to form a one-to-one relationship on a more intense level than he has experienced previously. Because of his group experience, however, he can withdraw, when the transference becomes too intense, into the safety of the group, only to be forced into the intense situation once again when the next individual session arises. This safety valve of the group will frequently afford enough protection to the patient to allow him to stay in therapy and maintain his transference to the therapist. Otherwise, he might experience intense withdrawal in twice-a-week sessions in individual therapy.

Because of the severe psychopathology that lies behind the drug abuser and his family structure, no method in the therapist's armamentarium should be disregarded. Careful survey of each individual patient and his family should be made to evaluate the best mode or modes of treatment. As has been mentioned previously, it is usually important for the family to be in some form of therapeutic situation, separate from the child, in order to avoid the pitfall of having the child return to the pathological home structure. At times, family groups or individual family therapy is all that is required. At other times, this approach should be supplemented by parent group therapy, or individual therapy for two parents together or for each separately. Once the therapist has a thorough understanding of the patient, how he relates to his family and how the family relates in turn to him, the combination of therapies necessary for the treatment of the patient becomes apparent. If the therapist neglects the psychodynamics in any of these situations, the patient may leave therapy, return to his previous way of life, or even attempt suicide.

Those forms of group therapy in which the therapist takes an active part in expressing and drawing out feelings from the group are an important adjunct to the treatment of the drug abuser. These forms of therapy rarely can be used without a more analytic mode of therapy at the same time. In "active" therapy, feelings are intensely mobilized for the moment, but rarely maintained over a prolonged period. They do not sustain the depth of emotion on the infantile level that is necessary for a character change, but they do mobilize feelings that have long remained dormant in the psyche of a person with a behavioral disorder.

In the encounter method, each member is expected to share his feelings towards the other members of the group in an active and emotional way. In the method of a sensitivity group, a leader conducts various forms of "experiments," which the patient goes through to help him recognize his own fear, mistrust, love, or sexual desires. In a "primal scream" group, patients lie on the floor in an infantile position, screaming for the absent parent. Individuals in any one of these modes soon become aware of long-repressed desires and feelings. They bring these feelings to the surface far more quickly and more intensely than in traditional analytic approaches. The highly constricted attitude of many members is reversed as they learn to handle feelings that were previously thought to be dangerous. They learn that they can experience rage, passion, or love—without necessarily leaving themselves vulnerable to hurt.

The group leader in such therapies must be willing to guide the group actively in directions in which *he* wishes it to go, to try to bring out those feelings which he suspects to be beneath the surface. In the marathon technique, where a group may persist for twenty-four, forty-eight, or seventy-two hours without let up, the exhaustion that is felt will often fragment the ego sufficiently to let feelings through with slight encouragement from the therapist. Because of the intense involvement of the therapist, many of the oral needs of these individuals are fed concurrently. Because of the feeding of these needs the patient is able to develop intense feelings for the therapist, which the therapist can then continue to encourage.

The major drawback to the use of active groups alone is that they never allow for a replaying of patients' infantile needs as clearly as they do in the slower, more tedious, analytic group. Active therapy completely eliminates the use of the transference interpretations, which is the major therapeutic tool of the analytic therapist. It seldom allows interpersonal transferences to develop among other members of the group, to allow them to play out their infantile needs among one another.

Because of the recent popular upsurge in this type of group therapy, its benefits have often been overrated. One of the dangers of these groups is the very high expectation of both the patient and the therapist—without recognizing the group's limitations. The analytic group has severe limitations because of its inability to reach the individual with egosyntonic symptoms. The activity group has the limitation of not being able to reach the depth of feelings that is necessary to make the character change. A combination of the two groups, therefore, is often more beneficial than either one alone.

Early in treatment, the marathon, the encounter, the recon, or the sensitivity group is important to mobilize the feelings that have been buried beneath the surface. Symptoms are then turned from the external world, to the internal life of the patient, making them egodystonic. Afterwards, when the patient is aware of the pain that he is in, he can make use of an analytic type of group, which will enable him to delve into that pain and to recognize its infantile, unrealistic origins. Unfortunately, prejudices among therapeutic schools have often negated the possibility of taking advantage of both of these techniques. Each school jealously guards its sense of therapeutic integrity and is fearful of incorporating any other therapeutic technique—almost a model on the professional level of sibling rivalry.

—14—
Concept Therapy

A third general classification of therapies may be distinguished not by the number of people involved but by the behavioral or psychological concept on which they are based. Both one-to-one psychotherapy and various forms of group therapy can obviously play roles in any of the treatments here discussed.

Alcoholics Anonymous is well known for its reliance on the concept of interpersonal help: people suffering similar emotional problems support each other. In the 1950s, Synanon burst on the scene with a similar idea applied to drug addicts. Other communities sprang up in imitation, some of which concentrated on symptoms rather than on the underlying psychological causes of drug abuse. For example, the latter groups employed the behavioral modification technique of reward and punishment. Punishments are "object lessons," ranging in severity from confessing one's errors in the community by wearing signs or having one's head shaved, to sitting in a "prospect" chair to ponder one's sins.

The basic rule of these new therapeutic communities is, "Thou shalt not use drugs." Members support each other in a sort of monastic life by sharing work, being available to "rap" with others when they need help, and otherwise "stuffing their feelings." They must accept the authority of staff members without question.

Therapeutic communities can be either full-time or part-time, in which the resident is trusted to sleep at home. The 24-hour community is obviously intended for the hardened drug abuser, who needs full-time support and protection against his addiction. One drawback is that the patient has no chance to face ordinary daily challenges, and must be evaluated for "reentry" to society by the staff. Group techniques and peer pressure similar to those of AA are used rather than individual counseling.

The full-time therapeutic community almost always makes some attempt to uncover the pathological causes of drug abuse, rather then concentrate only on symptoms. Group therapy techniques are popular, including encounter, marathon, and sensitivity sessions. Some communities encourage "re-con" groups, in which one member can vent his anger in a controlled setting against another member of the community.

The daily routine of a full-time member includes rising at an appointed hour, usually quite early, with the prospect of being "faced down" by the entire community if he backslides. He generally has an assignment in the kitchen or in maintenance, with a hierarchy of positions he can achieve as his functioning improves. During the course of the day he may employ his talents in entertainment, sports, reporting on the outside world, or in a variety of business tasks necessary to the survival of the community. In group meetings he uses peer pressure to reenforce those who are weakening and want to "split" or "get high." As we have seen, the psychodynamics of the addict make him highly sensitive to group influence.

Staff members are usually ex-addicts. The addict can identify with the feelings of the ex-addict and is better able to relate to him than to the schooled professional. They can also dedicate more time to their work—an unbelievable 24 hours-a-day, seven-days-a-week grind. The addict finds in the staff member who is an ex-addict an opportunity for a positive, close, and trusting relationship which he has been missing all his life. The ex-addict can also spot deceptions a professional would miss, and can be believed when he offers his help. For the first time, the addict is fulfilling his oral needs, and senses a situation in which he can be wholly dependent. The burden on the staff member who is a counselor is accordingly heavy. Any misstep on his part—a show of anger or any other antisocial behavior—can cause a serious conflict in the addict who has identified with him.

As he accepts the rigors of community life, the rules, and the fear of punishment, the addict finds a home and a new "family." He sees a caring and concerned authoritarian figure and a peer group intersted in his welfare. Primarily what transforms him from now on is the recognition, forced on him by his reliance on his counselor and peers, that he *is* dependent after all. His pseudo-independent facade falls away, and, though he may have come to the community only through pressure of the court or because of the panic of not having his drug available in the street, he unwittingly has undergone a change in his life style. In any other therapeutic situation, the decisive breaking of the addict's pseudo-independence is quite difficult to achieve.

Because it takes six to eight months to motivate an addict to this stage, and another six months perhaps to break down the "street" image which the addict uses to mask his unproductive, unmasculine idea of himself, therapy is understandably a long-term affair. In the second stage, the ridicule by the community of the "street" values of being a jailbird, shooting heroin, and "scoring" with girls is underlined by his work schedule and his "straight" clothes. The community begins to build an image to replace the old one. He is given more demanding tasks, such as that of a "rapper" for younger members of the community in need of help.

The fear of failing in his new responsibilities can sometimes create a will to fail—to avoid having to face more demanding tasks. The addict typically breaks a rule which drops him to a lower level each time he starts to succeed. "Object lessons" are used to break this pattern which force him into situations requiring responsibility until he realizes they represent no danger to him.

When he is ready for reentry, the now ex-addict may temporarily return for part of every day to gradually wean himself away from the community. Because of the dependency he has developed in treatment, he may employ any number of ruses to rejoin the community, such as falling back into drug use. In some communities, such as Synanon, the dependency on the community is encouraged and the ex-addict is always considered a member to be welcomed back for support at any time. In any case, parents for whom some dependency is still felt should also be counseled, so as to be able to adjust to the new dependency needs of the ex-addict.

The concept therapy is now complete; in most cases, individual or group therapy must now treat the deep-seated emotional problems that underlie the patient's drug abuse. In too many cases, the products of concept therapy are "robots," whose defenses are so severe that emotional expression and the chance of forming interpersonal relationships are sharply curbed. The preaddictive personality—insecure, inadequate, infantile—emerges, and drugs or alcohol again rush in to fill the gap. The failure rate among therapeutic community "graduates" has been high for this reason. This is a situation in which the professional world (of psychiatrists, for example) fails to recognize the contribution of the therapeutic communities in handling behavioral problems, and the concept therapists fail to recognize the contribution of psychotherapy in dealing with emotional problems. Both are necessary, and their cooperation is long overdue.

The part-time treatment center comes closer to combining the best of behavioral and emotional therapy. A resident of a day-care center must be able to return at night to a home that is not disruptive, or during the day to a job or to school. The cooperation of parents in both cases is paramount. Substituting a foster home or other residence away from home is generally impractical, for the child will break away and return even to a home filled with strife and turmoil, just to test his acceptance there.

Each day-care case must be treated individually, to minimize the anxiety that the child feels returning each night (or day) to the original source of his problem. Counselors, social workers, and psychologists must cooperate to attempt to change the personality of the child that was so susceptible to stress. The professional is needed to interpret and advise on the psychodynamics of the youngster's defense mechanisms, but an ex-addict staff is perhaps more important, as it is the only source for identification.

Many youthful drug abusers have seen a variety of psychotherapists, and have experienced only mistrust and failure. The day-care counselor must become a friend, an older brother or sister, and a parent substitute all at once. Most important, he or she must be prepared to be a model of behavior *outside* the center, where contact with patients frequently occurs and is always more sensitive to interpretation by patients. For this reason, the day-care counselor has a more demanding job than many staff members of a full-time community, which they can leave without concern for outside contacts with community members. Many day-care centers create the same aura of hostility that the child has seen at home when psychiatrists and social workers quarrel with counselors instead of recognizing the special need for each of their functions.

The role of the psychiatrist in a day-care center is to interpret the emotional disorders of patients in such a way that the staff can control the reactions which occur when their defense mechanisms are swept away. In a 24-hour community such control of symptoms is easy because the member is isolated from the objects of his frustration or anger. But when a child must return home each night and confront his parents, serious problems can result. The child begins to realize the rage he has for his parents. He must face this rage to throw off the self-destructiveness and passive-aggressiveness he has used as a substitute. In many cases, the child abuses or attacks his parents, if he is not counseled about the genesis of that rage. His aggression must be rechanneled, and the counselor must be aware that he cannot force the symptoms—as is done in a 24-hour community—without risk of driving the child away or even to suicide.

Because of the dependency needs of the members of a day-care center, 12 to 14 hours of supervision a day are usually necessary. Such a schedule, however, makes a split-shift feasible so that an evening program can be run simultaneously. The evening schedule becomes a reentry program for youngsters who can begin to go back to school or to jobs, and also an emergency station for day residents who may run into occasional problems at home.

As in a therapeutic community, the first step to therapy is the establishment of a dependency in the child on the center or the counselor. As time goes on, however, the counselor must wean him away from this dependency and so must have a sensitivity to the dependency-independency conflict of the resident and to the resident's defenses against this conflict. It becomes clear that behavioral modification alone, as practiced in a 24-hour community, is scarcely enough. But the group therapy techniques of that community work equally well in the day-care center.

It is in the precariousness of the child's behavior in a day-care center that there is a difference in a counselor's approach to him. The counselor must constantly be prepared for behavior that the child vents on him in lieu of the child's parent. The counselor must avoid criticizing the child while disapproving of the child's behavior, and must tolerate certain behavior, such as verbal abuse of the parent by the child, rather than rechannel these impulses into drug use or suicide gestures. At a 24-hour center, such situations can be controlled with disciplinary methods.

The child in the care of a counselor typically cannot separate his thoughts and feelings (that is, himself) from his actions. His infantile fantasy equates the former with the latter. So he feels guilt for his feelings as well as his actions. The counselor must separate the two, in effect by saying, "I don't approve of what you are doing, but that doesn't mean I don't approve of you."

Similarly, the counselor must distinguish his disapproval of a child's actions by the *reasons* for those actions. We have seen, for example, how the rage a drug abuser finally discovers he has for his parents can become frightening as it is externalized—and it must be externalized in the process of therapy. If the counselor feels threatened by this anger, especially if the patient is psychotic, he must be careful how he reacts. If he shows fear, the patient himself will become fearful of his own rage. If he stands firm against his anger by saying, "I won't permit this," then the patient's fear will be heightened by the thought that the counselor doesn't trust the patient to control his anger, but must control it himself. But if the counselor says, "I expect you to control yourself," then the burden is placed on the patient. In some cases, the resident may have a healthier ego structure and be less threatened by loss of control of his anger. The counselor can then allow it to develop. His rage begins to pose fewer threats. The subsequent alleviation of his depression opens the way for more assertive expression on his part.

Assertiveness is a general characteristic which might include "aggressivity"or aggressiveness in the sense of assuming responsibility. Vague though this concept may be, it is clear that the first step toward reestablishing a person's self-esteem and self-worth is the ability to assert oneself in thought and action. The drug abuser is often deterred from productive growth by fear of his own aggressivity. This assertiveness should be encouraged within certain verbal limits. For example, the therapist must tell the drug abuser, as he would tell the psychotic, "We know you can reserve those feelings for a better time." This is quite different from saying, "I think you ought to stop acting that way." In the latter case, the patient *incorporates* or swallows the counselor or therapist whole as an auxiliary superego and loses another chance to establish his own personality.

Assertiveness on the part of a day-care resident should be encouraged by offering him further positions of responsibility. His aggressivity is thus confined as a healthy reaction. If he misuses his new-found responsibility, care should be taken to make sure either that he is acting out hostile needs towards younger members, or simply that he finds the new responsibility threatening. If the latter is the case, the counselor must be careful not to reinforce the patient's feeling that he cannot assume responsibility. Therefore, in most cases the patient must be forced to assume his tasks, no matter how many times he tries to show he cannot handle them.

The "object lesson" has a role to play in the day-care center, but a different one from that of the 24-hour community. In the latter case, object lessons are a necessary device to ridicule and break down the unrealistic image of the hardened addict. The adolescent in a day-care center has no firm self-image to be broken. In the part-time community, too, object lessons must be quite specific. In the typical home environment, punishments are indiscriminate and arbitrary, so the day-care center must studiously avoid any reenactment of that scene.

Behind all of these techniques and counsels, the underlying importance of the peer group influences cannot be stressed too much. The counselor must do everything possible to foster the natural tendency of the drug abuser to rely on his peers. When so-called "negative contacts" show signs of occurring, the counselor must head them off. Self-destructiveness always remains a factor in the drug abuser's personality. When the resident appears headed in this direction, the counselor must interpret this aspect of his ambivalence. The goal of this type of therapy is to encourage positive aspects of his relationships, which will supply the interpersonal contacts which allow the final break from the community, and the assumption of his independence.

Counselors must discuss the trauma of the loss of a friendship formed in the community. In effect, the center must become a model for forming one-to-one relationships, with all of the care that must go into them in advance and the maturity that must follow an unexpected termination. The old virtues of loyalty, brotherhood, and solidarity must be encouraged in terms the adolescent can understand.

A strong third arm of the day-care center, psychiatric social work, is necessary to deal with parents and assure their cooperation. Parents of day-care patients require particular attention, since their daily interaction with their children is highly sensitive. The first concern of social workers is to *protect the patient or resident*, and not to work out the psychic needs of the parents. Parents may attempt to pit one staff member against another, making communication between counselors and social workers essential.

A common motive in parental antagonism is the reliance of one or both parents on the child as a way of acting out their own needs. Counselors must be wary of parents who insist that their children are uncontrollable, the therapy doesn't seem to be working, or the center's routine is too demanding. The parents may simply want to avoid any therapeutic change in their children because such change would work against them.

When parents are strongly against continuation of their child's therapy, a parental group may be used to bring peer pressure, or individual therapy may be called for. The center team must combine their thinking to make such a decision. Experimentation with various types of parental therapy is often the best procedure. Where drug counselors have been used to deal with parents, the routine is simplified but the age difference between parents and counselor leads to serious problems. The children lose their confidence in and identification with the counselor for "going over to the other side." Parents may sense a bias in the counselor in favor of the children, and so resist therapy for themselves. Parents may simply not trust a younger, apparently less experienced counselor. In general therefore, our experience has been that the simplicity of having a single counselor deal with an entire family has many more difficulties than a group approach by the center team.

The day-care center is a laboratory of personal behavior for parents and children alike. Its strength lies in its very fluidity. It can use almost the total range of therapeutic techniques, because it places almost every type of demand on parent and child. In a way, the day-care center requires therapeutic techniques of great artfulness for the simple reason that this situation most closely resembles life.

Various types of concept therapy can be combined, although there are obvious conflicts between outside, analytically oriented therapists and therapists in a 24-hour community. But where a drug abuser is also psychotic, he may require both therapies. He may find himself in a therapeutic community because of lack of hospital facilities for his circumstances. He may require antipsychotic medication, which demands the supervision and guidance of a psychiatrist. His prior self-medication for his psychosis may have been drugs, and without dealing directly with this psychological problem, no break can be made in his pattern of behavior.

We have already seen how the psychiatric professional can contribute his expertise during the reentry stage of concept therapy. This is where respect for each other's purposes is critical for the counselor and for the outside psychiatrist. The latter is working with behavioral defenses and internalized conflicts; he must have the full confidence of his patient. The counselor relies on behavioral modification, and may wish to have privileged information that is a matter of confidence. If the patient cannot rely on the confidentiality of his dealings with the psychiatrist, he will likewise be unable to profit from the strict regimen of the therapeutic community.

The cooperation of combination of therapies usually takes the form of deferring one to the other. In some cases, acting out becomes so self-destructive that the responsible psychiatrist must see the necessity for behavioral modification. In other cases, the 24-hour community must sense the psychopathology of certain of their residents and bring in professional help. In either case, the natural antipathy that exists between those with advanced degrees or many years of schooling and those with practical experience must be overcome for a successful combination of programs in the interests of drug abusers. Too often, professionals as well as untrained counselors become involved with their own needs in drug-abuse programs. Perhaps we should redefine these words. A professional is one who approaches an individual with that person's needs in mind, and no diploma can add or detract from his performance. The therapist who acts to serve himself acts unprofessionally, no matter how sacred his pedigree might be. For anyone who has worked with troubled young people, the only thing worse than the meddling of parents is the further meddling of those who call themselves healers, but cannot heal even themselves.

— Part Four —

The Public Problem

Let's have an end to public moralizing and find a way to the understanding of individual needs.

—15—
Why Drugs? Myths and Misinformation

In any group as diverse as that of drug users, there is bound to be a diversity of interests, of opinions, and of motivations for living that way of life. Let's step back for a minute and consider the whole drug scene from another vantage point. Our question is constantly "Why?"—for any clue we get to that question is enormously valuable in answering "What can we do about it?" Whatever repetition there is in the following approach is intentional; there are insights to be gained by turning the subject around, even though we seem to be using the same method of trying to discover the mechanisms which cause drug abuse.

Let's start first with the obvious observation that many users are simply *experimenters*. After coming into contact with a drug, they are willing to experience its effects once or twice, out of curiosity or a feeling that they "want to know what's going on." Following these experiences, they do not return to the drug again. They are aware of the dangers involved and are not willing to run the risk of prolonged usage.

An occasional user of marijuana and hashish is an individual who functions normally in his day-to-day living, attending school or work without interruption, forming wholesome relationships with members of both sexes. Businessmen, lawyers, doctors, and politicians often fall into this category. He is of no more danger to himself or to society than is the occasional social drinker. He is usually happy and self-contained, and feels a sense of fulfillment without artificial means. In contrast to these "stable" individuals, none of the following groups can be considered as being involved in a "normal" use of drugs, and fall into the category of the drug abuser.

The abuser has become psychologically dependent on one aspect or another of the drug or drugs that he is using. Most of these people do not limit themselves to one form of drug and there is an increasing number of mixed addictions. In this category are included the "weekend" hard drug user, the marijuana abuser, the continual hard drug abuser, and the addict.

The "weekend" drug abuser includes the type who uses drugs sporadically in social situations, and the one who rarely lets a weekend go by in which he is not "stoned." These people are not fully aware of the consequences of their drug use because they deny the probability of danger to themselves. They are not emotionally dependent on the drug itself but rather upon gratifying a self-destructive need. Their drug use is often accompanied by apathy, meaningless rebellion, paucity of friends and relationships, and an inability to function to their capacity in school or at work. There is a lack of outside interests and their most frequent activity is "hanging out." Even the college-bound student or the athlete who may be in this group is unlikely to be functioning near his peak. The use of a dangerous drug, then, is a symptom of an underlying problem accompanied by a neurotic need—as we have discussed previously.

Despite the fact that marijuana does not cause physical dependency, there is usually an emotional dependency on the drug. Contrary to many opinions, it is not the *drug* that causes dependency but the needs of the individual. It is the presence of the drug that allows the individual to gratify his dependency needs. The marijuana abuser can vary his use from three or four joints a week to three or four joints a day. He finds that without the drug his life is anxiety-ridden. He has difficulty coping with day-to-day tasks. Socializing becomes uncomfortable, as he feels inadequate with his peers. He uses the drug to relax and feel capable of functioning. School is no longer important. His family and old friends become alienated and he finds himself in a new life style.

The marijuana user differs from the alcoholic in some ways, but is similar in many more. When an alcoholic sneaks a drink he is ashamed of his problem; the marijuana user is proud of his accomplishment and often boasts among his peers of the amount he has smoked. The alcoholic is aware that with increasing amounts his faculties diminish. The marijuana user believes that his perception is heightened, that his ability to perform increases, and that overall he functions better than before. There is evidence to show that when used in small amounts by people who are used to its effects,marijuana does produce such desired results. This does not hold true, however, when the amount increases and the person begins using it as an escape from day-to-day life.

Hard drug users are those who take drugs on a frequent to daily basis without becoming addicted. There is a strong psychological dependency on the drug and functioning without it becomes extremely difficult. The pressures of school or work become intolerable. They begin to function only when under the influence of the drug. Their familial relationships have generally shattered. There is a constant state of tension between themselves and their parents or other authority figures. Siblings who do not use drugs are generally alienated. Peer and sexual relationships are definitely drug oriented, and there are few "straight" friends in their crowd. Sexual acting out is a frequent accompaniment of their drug abuse, which increases the need for the drug to alleviate the sense of guilt and shame that attaches to sex. Love relationships are infantile and filled with a dependency need. It becomes a spiraling downward cycle as the loss of pride and shame increases with the lack of functioning. Before any change can occur, the cycle must be broken.

Cells in the body of the addict have altered to a state where they require the drug in order to function. The withdrawal of the drug leads to severe physical symptoms. Newborn infants of addicted mothers are addicted to the drug and go through a withdrawal period right after birth. The addict's whole life soon becomes involved with the pursuit of obtaining the drug. The violent crimes of the addict are usually an attempt to obtain money to obtain the drug, or are performed out of a sense of fury and frustration when his needs are not met.

A highly infantile emotional pattern underlies almost all drug addicts. Their tolerance for pain and frustration is almost nonexistent. There is no pride, no self-respect, and no self-esteem. Life becomes an almost animal-like existence, in that there is nothing left but the pursuit of the basic necessities for survival.

The human values of intellect and aesthetics disappear; loyalty and true friendship are rare.

All controversial subjects are filled with legends and myths, but those subjects that have remained underground and taboo have more than their share of misconceptions. Often one hears tales of the terrible "dope fiends," innocent young men and women who after a first shot of heroin are turned into vile rapists and murderers. Moralists still refer to marijuana as "the deadly weed" that totally distorts the senses and transforms the user immediately into some beast capable of committing crimes he would never think of ordinarily. There are an equal number of misconceptions among drug users, too, concerning the harmlessness of drugs. Misinformation from supposedly knowledgeable sources has confused the drug issue for so many years that much of the public is still ignorant of what is really happening.

There is no such thing as a drug "fiend." When well-bred children take a shot of heroin the reaction is very similar to that of not so well-bred children. They become relaxed and quiet and generally "go on the nod." It is the addict who is looking for money to buy heroin who is the danger to society. His desperation and frustration are what lead to violence.

A recent mistaken concept about the opiates involves methadone. It is widely believed that one of the advantages of using methadone in the treatment of the heroin addict is that it does not cause a "high" as do the other opiate derivaties. This is a half-truth. Methadone, in fact, does give the individual a feeling of well-being and comfort. What is lacking is the initial "rush" that is achieved with heroin. Nonetheless, despite the lack of this rush more and more people are beginning to turn to methadone as the primary drug of choice because of its availability and low cost.

Aside from the opiates, more misinformation surrounds the use of marijuana than any of the other drugs. Some of the generalizations people have made about marijuana come from a rare, isolated experience of one who is emotionally unstable prior to the use of marijuana. The behavior that was attributed to the drug might have manifested itself without marijuana as a stimulus. No doubt marijuana lowers inhibitions in much the same way that alcohol does. We are well aware that people who drink do not behave the way they would under ordinary circumstances. The main danger from marijuana, like alcohol, is that when a person undertakes a task that can be dangerous, the loss of perception caused by these drugs can make him a menace to society.

The vast majority of marijuana users never go beyond the marijuana phase. There have been estimates recently that 70–80 percent of college students smoke marijuana. It is well known that many business executives, doctors, lawyers, politicians, and police officials indulge in marijuana smoking. Most of these people are basically stable and contain enough ego strength not to become totally involved in drug use. The person who goes from marijuana to heroin has a preaddicted personality. He lacks ego strength; he is highly dependent, with little self-esteem or self-worth and would become a heroin addict had he never tried marijuana.

Most of these myths develop out of a moralistic climate that has been present since the 1920s. Similar myths grew about the use of alcohol prior to prohibition. Now, as then, these beliefs form the backbone of the arguments of those who are opposed to any leniency where the use of drugs is concerned. These arguments are not based upon sound scientific findings but rather upon generalizations taken from specific incidents. There is no evidence, statistical or otherwise, to back up their all-or-nothing thinking.

There are an equal number of myths that have grown up in the drug culture that are just as fallacious. Perhaps they are more dangerous because they allow the individual to indulge in activities that are basically harmful to himself without realizing the danger that is involved. There are two primary misconceptions about the use of marijuana. First, many individuals believe that it is a safe drug. They will quote study after study showing that no one has proven that marijuana is dangerous. But they ignore the basic fact: marijuana is dangerous in the same way that alcohol is dangerous—in critical situations where perception and judgment are necessary. *It is more dangerous, in this sense, than alcohol because the person under the influence of marijuana refuses to acknowledge the perception distortion and argues that he functions equally well with or without the drug.* There is danger, too, that one may not be an individual who can tolerate the drug. Many people develop anywhere from mild to moderate and even severe paranoid reactions under the influence of marijuana. The user then answers, "Well, of course, this will never happen to me," which is like a spin of the cylinder in Russian roulette. It is possible for an individual who has smoked on numerous occasions to experience a particularly bad incident and a psychotic reaction.

I first became aware of this situation while I was a staff psychiatrist at Hillside Hospital. A young man entered the hospital in a severe paranoid state. He was constantly frightened of individuals and could not form friendships because he was afraid that people were out to harm him. He was in a constant state of anxiety, which caused him to be unable to function for an entire year. Prior to his admission to Hillside he had been to Bellevue and St. Vincent's for a total of four hospitalizations of about three months. He was a graduate student at a university and was majoring in business administration. He graduated from college summa cum laude. Though emotional problems were manifest prior to his paranoid state, none of them reached the severity that required either hospitalization or psychiatric intervention, nor did it ever interfere with his functioning. He was active on campus both academically and politically as one of the organizers of a conservative youth organization. This showed him to be of a conservative nature, a philosophy which followed throughout his history.

He had smoked marijuana on two or three occasions with no apparent effect, until one night when he smoked with his roommate. He was a slightly built, 160-pound individual; his roommate was an ex-football player who weighed about 230 pounds. He became totally irrational. He made overt homosexual advances to his roommate. Then he tried to get into a fight and beat him up. He then threatened to jump out of the window of the thirteenth-floor apartment, and only the strength of his roommate kept him from following through with the act. He was finally physically subdued. The police and his father were called and he was brought to Bellevue Hospital for the first of his numerous admissions. It took a year and a half of hospitalization and therapy to allow this individual to return to functioning at his original level. It is true that many of the actions which became overt that night and on subsequent occasions had been lying dormant. However, it is improbable that these feelings would have emerged so abruptly and with such destructive effect without the use of marijuana.

This is a striking example of the danger of marijuana; certainly it does not occur in the vast majority of cases. Anyone, though, who is familiar with people who smoke marijuana is familiar with stories of people becoming paranoid to some degree. They become frightened that there are plots against them and that even their friends and family constitute a danger.

One of the "facts" heard about marijuana is that it improves one's performance. Despite studies that show that it is true in certain instances, as mentioned earlier, for a vast number of users of marijuana it is a blatant rationalization. Many youngsters tell me that they drive cars while under the influence of marijuana and nothing has happened; in fact, they say their reaction time is even better than before. There are numerous studies to show that reaction time is markedly slowed under the influence of the drug and that the risk of an accident is increased. What marijuana is actually doing in these cases is alleviating some of the anxieties that go along with the task. When anxiety is diminished, functioning will increase; however, the side effects of perceptual distortion generally override the desirable effects of the anxiety diminution.

If one is in doubt, it might be convincing to make a video tape recording of oneself performing under the influence of marijuana, and play that tape back when one is straight. When this is done the individual is usually surprisingly disappointed with his performance, despite the prior feeling that the performance was better than usual. The feeling and the memory of the superiority of the performance comes from the distortion that takes place with the drug. It is this pleasurable perceptual distortion that one seeks from marijuana.

Recently, there has been an increase in the use of quaaludes and sopors which are pharmaceutical names for methaqualone. The dramatic increase in the use of this drug is no doubt due to the legend that these drugs are relatively safe. Recent studies have shown that the methaqualone drugs, though less addicting, can manifest the same addictive symptomatic effect as do the barbiturates. Withdrawal can lead to convulsions and death.

In the 1950s Timothy Leary, then a Harvard professor, began

to advocate the use of LSD. He developed a substantial following among adolescents and some "intelligentsia." He expounded the theory that LSD possessed a mind-expanding quality that allowed a deeper and more insightful understanding of oneself and the world than would be possible without the drugs or with other drugs. The proponents of the new ethos that grew up around LSD looked back to the times when the Indians in the Southwest used peyote and mescaline in order to improve communication between their gods and themselves.

The use of hallucinogenic substances took on all the fervor of a religion, including many pagan trappings. Each trip was looked upon as a new mystic experience, in some ways in direct communication with "God." Users consistently denied the dangers of the drugs, and, as they became more and more involved, their mental faculties deteriorated to a point where they no longer had the power to decide whether or not the drug was dangerous.

Far from being a mind expander, LSD really constricts the thinking ability and the insight of an individual. It causes an organically induced psychosis that prevents the individual from understanding reality. If one were to visit the back wards of a state hospital and interview a patient who is *not* taking phenothiazine drugs, one would hear the same type of marvelous revelations about himself and the world—revelations based on a kernel of truth that seem very real to the individual. Some of the insights are interesting, and it would require an analyst years of working with a patient on the couch to recognize the same truths. As with the psychotic, however, the LSD user does not fully understand the meaning of his insights. Mind expansion is best accomplished by having a wondering and inquiring mind, looking deeply into subjects and ideas, and not taking things at face value.

The drug is also dangerous, as psychosis has led to suicide and accidental deaths. It has led to nonpremeditated, drug-induced psychotic murders, often of infants. In less dramatic, but equally tragic cases, individuals have remained psychotic for many years and in some cases even permanently.

The discussion between the drug user and the nondrug user often centers on whether drugs are good or bad. A drug is neither good nor bad; it just "is." You cannot apply a moral value to an inanimate object with no will of its own. A drug can be classified, however, as dangerous or safe. A person can be classified as intelligent or nonintelligent if he does a smart thing or a stupid thing. In most instances it is stupid, and not bad, to use drugs. Once the morality issue is eliminated from the drug discussion, then, perhaps, clear thinking will replace the emotionalism that has been the curse of the subject.

—16—
Finally, The Family

I was once asked to discuss drug prevention and treatment with mayoral aides in New York City, and came away staggered by the misinformation and archaic concepts which pervade the city halls of our nation. On one hand, legislators and city officials call for harsh punitive measures, which make no distinction between the hardened professional, who is in drugs for profit, and the teenage victim, who is trapped by his own spiraling needs. On the other hand, well-meaning public agencies experiment with a wide variety of treatment, which unfortunately consists mainly of lecturing and preaching.

The therapeutic program presented in this book is no easy solution. It is based on a strict regimen, quite different from classical psychotherapy, requiring the cooperation of a team: a psychiatrist (or psychologist or psychiatric social worker familiar with the mechanisms as described in the previous chapters), an ex-addict who has experienced the value of total-involvement treatment, and a social worker who can effect the changes in the family that are necessary for the survival of the child.

The majority of children caught up in the drug and alcohol crisis are not addicts. For this reason they must be treated differently, though they can be helped by the ex-addict. They can also be helped by the cooperation of the school, by the intelligent use of law enforcement, and by the firmness of an understanding court. Finally, however, the burden falls on the family. *It is part of the public problem to make this responsibility clear.*

The school must take an active role in calling a potential drug problem to the attention of a child's parents. If a teacher notices a marked change in grades, discipline, or truancy, parents should be notified immediately. Sending notes and report cards home with a student, or even by mail, is frequently useless; the student will often destroy or alter the notes or intercept the mail. The teacher, drug counselor, or guidance counselor should contact the parents by telephone or call them in for a conference. Many parents cannot be bothered or plead lack of time for a conference, attitudes which indicate a negligence which may be at the root of the problem. In such cases, it is usually better for all concerned to drop the child from school if the parents do not cooperate.

The influence that drug abusers have on one another, as we have seen in an analysis of group pressures, is a strong argument for expelling obvious drug abusers. Schools have an obligation to protect their students from each other, and even from negligent parents The absence of parental cooperation must be viewed with the greatest concern by school counselors.

Unfortunately, many guidance counselors in high schools were trained for career-advice, and are out of touch with modern youth because of their age or professional background. Counselors with specific responsibility for drug treatment should be hired, even on a part-time basis, from therapeutic communities of day-care centers. In the present financial circumstances of most public schools, however, this alternative is far from easy to put into practice. Here again we come face to face with a public problem.

At the law enforcement level, it is perhaps even more difficult to overcome traditional responses to problems. However, farsighted courts and police departments are becoming increasingly aware of the emotional nature of an adolescent's drug abuse. Where therapeutic communities are available, judges are using their discretion to force parents and their children into some form of therapy, instead of punishing the misdemeanor or minor felony that brought them to court. More and more officers are being trained to overcome the frustrating indifference of many parents and judges, and they are looking beyond the family's home, neighborhood, or social status in deciding when to call a drug problem to the attention of the parents.

As the drug problem in most communities has tended to go underground in recent years, it has become increasingly important for parents to learn the behavioral symptoms that are not so easily masked. When the obvious physical signs of drinking or taking drugs are being disguised, these are some of the things parents must look for:

- a sudden and continuing drop in school grades
- class-cutting and truancy without apparent reason or shame
- a marked mood change unrelated to normal adolescent behavior
- constant depression, extreme belligerence towards parents
- a change in friends to those his parents disapprove of
- a sudden refusal to bring his new friends home
- a drifting away of older friends and neighbors

Naturally the parent must weigh these warning signs carefully before rushing into action, but it is usually the case that parents believe anything rather than that there is trouble looming. As we have seen, the worst reaction is anger and direct threats. There are two general guidelines for parents whose suspicions of drug or alcohol abuse are strongly aroused. First, the parent must shoulder responsibility early and decisively, and that means admitting the problem and realizing that it cannot be considered the child's problem alone. The child should be questioned without hesitation and with a good deal of skepticism. In general, the parent can follow the guidelines which we have seen for the therapist. But, secondly, the parent must not assume the role of a therapist. It is easy to say, "Seek professional advice." It is critical to know when and where.

It has been the whole purpose of this book to give answers to those two questions. The answers cannot be formulated in one-two-three fashion. But if the parent understands the dynamics of drug abuse as viewed from several aspects—from causes to symptoms to therapy—he or she can make an informed decision about when to seek treatment.

But what about *prevention*? No parent expects his infant to grow up to be a drug addict or an alcoholic. Indeed, it has become fashionable to deny any strong connection between the nurturing of a baby and his chances of growing to healthy maturity. We have been told by implication if not by explicit argumentation that the family may have been necessary in another era—a nonindustrialized, rural society—but now it has lost its usefulness or desirability. Often this is a viewpoint that is argued with honesty and sociological persuasiveness. More often, it is only a convenient rationalization for easy divorce and the abandonment of the roles of motherhood and fatherhood.

Yet from the standpoint of drug abuse alone, the formative role of the family can be ignored only by those who do not want to see. There is no better way to end this discussion of adolescent drug abuse than by reviewing the family mechanisms and structure that underlie the development of a child. If the prospect of having a baby and "raising a family" is only a distant dream (or nightmare!), nevertheless now is the time to become aware of how a child is molded from his earliest years.

The family is the primary source of the personality, identity, and individuality of a child. He learns his relationships with himself and with others through reflections of his relationships with his family. Even where grandparents live with the family, the parents play by far the primary role in influencing their child.

Contact with the mother is of greatest importance to the newborn. Even before the infant leaves the hospital, certain communication has been established between mother and child. Early maternal feelings are transmitted through body tensions; a mother who is tense in nursing is likely to cause tension and possible feeding problems in her baby. If she is relaxed, her baby will feel comfort and security in her caress. It has been widely accepted that the absence of loving physical contact between mother and child can cause depression and even death of an infant.

The baby is aware of a mothering figure, not necessarily a specific mother. An adolescent who is disturbed may revert to this condition in not being able to develop any consistency toward people to whom he turns for security. He will accept any source of security, without differentiating other qualities.

Within his first few months of life, the baby begins to separate his mother from others who enter his life. The mother is greeted by a smile of recognition and a reaching out. From the maternal figure he sees, the baby develops his idea of how humanity will treat him. The child's anticipation of future relationships is indelibly colored by the warmth and love, or the distrust and hostility, he finds at this age in his mother.

The importance of the father is perhaps harder to accept. Yet it is psychologically sound that a healthy child development is almost impossible without the presence of both parents. The father soon becomes the secondary object of the baby's love and affection. A girl's future relationships with men are based on what she has seen in her father; a boy's love for and identification with his father make it possible for him to resolve his early feelings for his mother. The father is usually seen as a figure of authority, the one who sets limits and gives advice.

In analytic situations, children frequently express resentment for their father if they, the fathers, are passive and dependent while their mothers are strong and authoritarian. Such roles may be culturally assigned, but there are strong indications of an innate basis for them. Consider that the parent who first nurtures the baby is almost invariably the mother; if she later assumes an authoritarian role, the child becomes confused. The father is traditionally viewed as the one who stands between the household and the environment; if he does not live up to his idealized image of the one who assumes control in threatening situations, the children of both sexes show resentment. These roles are established, if not by feeding patterns, surely by the physiology of birth.

The family soon begins to offer something more than security and warmth to the child: it offers scope for exploration. In a healthy situation the child begins to go in his own direction and then pulls back for security (in exaggerated form, this struggle can become the source for serious conflict in later life). The straining for individual growth is openly exhibited in what are called the "terrible twos."

The rebellious, self-willed behavior characteristic of this age is a healthy outlet and a necessary part of growth. A totally compliant child never has the opportunity to test the bonds of parental affection. He will never learn that a parent can disapprove of his actions without denying love to the child. So the child can easily develop a fear of being abandoned that will last throughout his lifetime.

When the child first enters school, the anxiety of leaving the mother may cause what appears to be a fear of school, and in severe cases a fear of somehow destroying the parents through the child's ambivalent feelings. Separation at school can mean loss of love to the child; at the same time, the child expects nothing but love. A child may therefore wish to remain at home to avoid that dreaded catastrophe.

The delicate balance between limit-setting and love by the parents was clearly expressed by a young girl who told us, "It is hard to leave home when you don't know your parents will still love you when you return. It is much easier to leave home when you know that when you come back you will be welcomed with open arms." The conflict between dependency needs and independence goes on well into adult life.

At the preadolescent level, the conflict is subtle. The young child who is aware his parent or parents are home will easily run out to play after school. But if the parent is gone for any length of time, the child may become distressed, even though he does not appear anxious when he is physically out of sight of the parent. A superficial acceptance of the parent's absence can lead to the pseudoindependent characteristics we have seen in the drug abuser.

It is the availability rather than the actual presence of the parent that gives the child a sense of security in an alien world. For this reason it is important for a mother who must leave the house for any length of time to let the child know where she has gone. Even into the adolescent stage, the fantasy of the parent's availability is critical at the same time that the child begins to show rebelliousness and willingness to experiment with peers as replacements for his family.

The adolescent's rebellion is a necessary part of his growth toward maturity, yet is often mistaken by parents as rejection. The parents become anxious in their fear of drugs and sexual behavior. The apparent lengthening of the adolescent period because of the demands of modern society intensifies this problem. In previous eras, or even in previous generations, sons and daughters assumed roles earlier in life, and so were able to end their dependency perhaps five years earlier than now. Young men and women nowadays delay a career decision into college or graduate school which in effect lengthens their adolescence. Accordingly, the parental role is prolonged. The adolescent who is ready physiologically for exploration and sexual discovery is today being kept in a state of socioeconomic childhood. The pressure on parents to handle this new familial condition is unmatched in recorded history.

If parents are understanding of their child, yet firm in their setting of limits, if they love their child but are willing to let him grow in his individual way, can that child become a drug abuser? This is another way of viewing the major proposition of this book. The answer is yes, he can. But he can also be treated far more simply and positively. His abuse of drugs may well be the result of normal conflicts, and may be self-correcting.

Neurotic conflicts are a normal outcome of the development of a child. They begin with the first pangs of hunger that are not immediately allayed. When the child is toilet-trained, he feels a conflict between his mother's wishes and his bodily functions. At an age of three or four, oedipal conflicts between the child and the parent of the same sex develop toward the other parent. By the age of ten, such conflicts should be resolved, though they recur occasionally in adolescence. External factors, such as the birth of a brother or sister, illness, or school experiences, can add to neurotic conflicts at any age.

Drugs will offer themselves as a form of relief from any of these conflicts, in competition with healthy defense mechanisms. Yet the essential difference is this: drugs are here used only to relieve an anxiety without a self-destructive, antisocial component. Underneath is a healthy ego. *The basic signs of the vicious circle of drug abuse are severe depression, a self-destructive urge, and a resentment of parents.* This triumvirate generally cannot exist without a disturbed home.

We end where we began. The family relationship is at the heart of the problem of drug abuse. The family relationship means nothing more, or less, than the psychological ties of mother and father to son or daughter.

BIBLIOGRAPHY

Books

1. Abraham, Karl. *On Character and Libido Development*. W. W. Norton & Co., New York, 1966.
2. Ackerman, Nathan. *The Psychodynamics of Family Life*. Basic Books., New York, 1958.
3. Aichhorn, August. *Wayward Youth*. Viking Press, New York, 1935.
4. Arlow, Jacob and Brenner, Charles. *Psychoanalytic Concepts and the Structural Theory*. I.U.P., New York, 1964.
5. Berkovitz, Irving H. (ed.). *Adolescents Grow in Groups*. Brunner/Mazel, Inc., New York, 1972.
6. Brenner, Charles. *An Elementary Textbook of Psychoanalysis*. (Revised edition). I.U.P., New York, 1955.
7. Brill, Leon and Lieberman, Louis. *Major Modalities in the Treatment of Drug Abuse*. Behavioral Publications, New York, 1972.
8. Brill, Leon. *The Treatment of Drug Abuse*. The American Psychiatric Association, 1972.
9. Eissler, K. R., M.D. *Searchlights on Delinquency*. I.U.P., New York, 1949.
10. Erikson, Erik. *Childhood and Society*. W. W. Norton & Co., New York, 1950.
11. Feinstein, Sherman C., Giovacchini, Peter, and Miller, Arthur A. (eds.). *Adolescent Psychiatry*, Vol. I, Basic Books, Inc., New York, 1971.
    Freedman, Daniel. "On the Use and Abuse of LSD."
    Neff, Leonard, "Chemicals and Their Effects on the Adolescent Ego."
    Borowitz, Gene H. "Character Disorders in Childhood and Adolescence."
    Rose, Gilbert. "Maternal Control, Superego Formation and Identity."
12. Fenichel, Otto. *Collected Papers*. "Neurotic Acting Out," pp. 296–304. W. W. Norton & Co., New York, 1954.
13. Freud, Anna. *The Writings of Anna Freud*, Vol. VII, 1966–1970. Problems of Psychoanalytic Training, Diagnosis and Techniques of Therapy. I.U.P., New York, 1971.
    Acting Out (1968).
    Adolescence as a Developmental Disturbance (1969).
14. Freud, Anna. *The Ego and the Mechanisms of Defense* (Revised Edition). I.U.P., New York, 1966.
15. Freud, Sigmund. *The Complete Psychological Works of Sigmund Freud* Standard Edition. The Hogarth Press, Ltd., London, 1966 Edition.
    Vol. IX. *Obsessive Actions and Religious Practices* (1907).
    Vol. XII. *The Dynamics of Transference* (1912).

Vol. XIII. *Totem and Taboo* (1913).
Vol. XVI. *General Theory of Neurosis* (1917).
Vol. XIX. *The Ego and the Id* (1923).
Infantile Genital Organization: An Interpolation into the Theory of Sexuality. (1923).
Neurosis and Psychosis. (1924).
The Economic Problem of Masochism. (1924).
The Dissolution of the Oedipus Complex. (1924).

16. Freud, Sigmund. Collected Papers, Vol. IV. Basic Books, New York, 1959.
The Unconscious. (1915).
Mourning and Melancholia. (1917).
17. Fromm-Reichmann, Freida. *Principles of Intensive Psychotherapy*. University of Chicago Press, Chicago, 1950.
18. Glasscote, Raymond; Sussex, James; Jaffe, Jerome; Ball, John; Brill, Leon. *The Treatment of Drug Abuse*. The American Psychiatric Association, 1972. Washington D.C.
19. Goldberg, Carl. *Encounter Group Sensitivity Training Experience*. Science House, Inc., New York, 1970.
20. Greenacre, Phyllis, M.D. (ed.) *Affective Disorders*. I.U.P., New York, 1953.
Bibring, Edward, M.D. "The Mechanism of Depression."
Zetzel, Elizabeth Rosenberg, M.D. "The Depressive Position."
21. Grinspoon, Lester, M.D. *Marijuana Reconsidered*. Harvard University Press, Cambridge, 1971.
22. Howells, John G. (ed.) *Modern Perspectives in Adolescent Psychiatry*. Brunner/Mazel, New York, 1971.
23. Jacobson, Edith. *The Self and the Object World.* I.U.P., New York, 1964.
24. Lowenstein, Rudolph. (ed.) *Drives, Affects, Behavior*. I.U.P., New York, 1953.
Knight, Robert P. "Borderline States."
Schur, Max. "The Ego in Anxiety."
Spitz, Rene. "Aggression: Its Role in the Establishment of Object Relation."
De Groot, Jeanne Lampe. "Depression and Aggression."
25. Reich, Annie. *Psychoanalytic Contributions*. I.U.P., New York, 1973.
"On Countertransference." 1951.
"Early Identifications As Archaic Elements in the Superego." 1954.
"A Special Variation of Technique." 1958.
"Pathologic Forms of Self Esteem Regulation." 1960.
26. Rogers, Carl. *On Encounter Groups*. Harper & Row, New York, 1973.
27. Slauson, S. R., M.D. *The Practice of Group Therapy*. I.U.P., New York, 1947.

28. Schur, Max (ed.) *Drives, Affects, Behavior*, Vol. II, I.U.P., New York, 1965.
    Nacht, S. "Interference Between Transference and Countertransference."
    Jacobson, Edith. "The Return of the Lost Parent."
    Zetzel, Elizabeth R. "Depression and the Incapacity to Bear It."
29. Tarachow, Sidney, M.D. *An Introduction to Psychothearpy.* I.U.P., New York, 1963.

Articles

1. Beaubrun, Michael H., M.D., and Knight, Frank, M.D. "Psychiatric Assessment of Thirty Chronic Users of Cannibis and Matched Controls." *The American Journal of Psychiatry*, **132** (3), March, 1973.
2. Bieber, Irving. "On Behavior Therapy, A Critique." *Journal of the Academy of Psychoanalysis*, **1** (1), 1973. John Wiley & Sons, Inc.
3. Chessick, Richard D. "The 'Pharmacogenic Orgasm' in the Drug Addict." Archives General Psychiatry III., 1960, pp. 545–556.
4. Farnsworth, Dana L., M.D. "Dilemma of the Adolescent in a Changing Society." *Psychiatry Annals*, **3** (5), May, 1973.
5. Farnsworth, Dana L., M.D. "Toward a Theory of Drug Control." Psychiatry Annals, Vols. 3 and 4, April, 1973.
6. Frankl, Viktor E. "Encounter: The Concept and Its Vulgarizations." *Psychiatry Annals*, Vols. 3 and 4, April, 1973.
7. Greenbaum, Esther. "Unfolding of Identifications: Repetition and Change in Identity Formation." *Journal of American Academy of Psychoanalysis*, **1** (2), 1973. John Wiley & Sons, Inc.
8. Hochman, Joel Simon, M.D., and Brill, Norman, M.D. "Chronic Marijuana Use and Social Adaptation." (ibid. ref. 5).
9. Janus, Samuel, Ph.D. and Bess, Barbara, M.D. "Drug Abuse, Sexual Attitudes, Political Radicalization and Religious Practices of College Seniors and Public School Teachers." *The American Journal of Psychiatry*, **130,** (2), February, 1973.
10. Klonoff, Harry; Marcus, Anthony and Low, Morton. "Neuropsychological Effects of Marijuana." *Canadian Medical Association Journal*, **108:** 150–156, 165, 1973.
11. Report of National Commission on Marijuana and Drug Abuse. (1972).
12. Savitt, Robert A. "Psychoanalytic Studies of Addiction: Ego Structure in Narcotic Addictions." *Psychoanalytic Quarterly*, Vol. 23, 1963.
13. Zinberg, Norina E. "Heroin Use in Viet-Nam and the United States." *Archives of General Psychiatry*, **26:** 486, 1972.

Index